A Catalogue of Ecosystem Services in Slovakia

Peter Mederly • Ján Černecký
Editors

A Catalogue of Ecosystem Services in Slovakia

Benefits to Society

Springer

Editors
Peter Mederly
Faculty of Natural Sciences, Constantine the Philosopher University in Nitra
Nitra, Slovakia

Ján Černecký
Faculty of Natural Sciences, Constantine the Philosopher University in Nitra
Nitra, Slovakia

State Nature Conservancy of the Slovak Republic
Banská Bystrica, Slovakia

Institute of Landscape Ecology of the Slovak Academy of Sciences
Bratislava, Slovakia

ISBN 978-3-030-46507-0 ISBN 978-3-030-46508-7 (eBook)
https://doi.org/10.1007/978-3-030-46508-7

This Springer imprint is published by the registered company Springer Nature Switzerland AG
The registered company address is: Gewerbestrasse 11, 6330 Cham, Switzerland

Preface

Nature and the environment represent a key and unquestionable value in terms of human existence – perhaps the value increasing in importance the more people live in the world and the more humans influence and change their environment through their activities. The world community and the European Union are becoming more and more aware of this fact, with various corresponding international activities and policy initiatives. The first decade of the new millennium was very specific with global research and environmental assessment focusing on biodiversity and natural resources, resulting in the outcomes of a large-scale Millennium Ecosystem Assessment (2005) project, which raises the issues of biodiversity, natural capital, and ecosystem services. Another key document is the Convention on Biological Diversity (CBD) and its Strategic Plan. At the same time, the 2011–2020 period was also declared the *Decade of Biodiversity* by the United Nations (UN). The European Union (EU) has also adopted a number of important documents in the recent period – with the EU Biodiversity Strategy by 2020, which was adopted in 2011 and is worth highlighting. According to this document, the vision of EU policy is the protection, valuation, and adequate restoration of biodiversity and ecosystem services (natural capital) that ecosystems and nature provide.

Ecosystem services (ESs) thus represent a relatively new scientific and applied concept focused on assessing and valuating the importance of natural resources, ecosystems and their functions, and, ultimately, the services which nature provides to people not only in particular territories but also on a global scale. The simplest ES definition from the TEEB study says that *ecosystem services include all direct and indirect contributions of ecosystems to human well-being*. Although the ES concept appeared already sometime after 1980, it has received a higher rate of attention only in the last 20 years, with the attention growing almost exponentially in the last decade. Greater publicity for the concept was promoted by a study by Costanza et al. in 1997, which quantified the economic value of world ecosystems. The interconnection with the political scope was facilitated by publishing the results of the above-mentioned Millennium Ecosystem Assessment (MEA) project and other important projects and activities (e.g., TEEB – The Economics of Ecosystems and Biodiversity, IPBES – Intergovernmental Platform for Biodiversity and Ecosystem

Services, MAES – Mapping and Assessment of Ecosystems and their Services, ESP – Ecosystem Services Partnership, and others). As Costanza et al. stated in 2017, the ES issue has progressively crossed the scope of natural sciences, penetrating into the practical and political area, and gradually also into the field of economics – but the level of its practical application is still not sufficient enough.

The updated National Biodiversity Protection Strategy for 2012–2020 was also adopted by the Slovak Republic, which committed itself to meet certain objectives – the mapping and assessing ecosystems and their services is one of them. The decade is slowly coming to an end and it will probably be necessary to admit that several objectives have not been met. Slovakia is also lagging behind in the assessment of ecosystems and their services (the MAES process) compared to other European countries. Although the Ministry of Environment, in particular through its professional organizations (SNC SR – State Nature Conservancy of the Slovak Republic, SEA – Slovak Environment Agency), supports and implements certain activities, the comprehensive assessment of the ES has still not been achieved. That is also why this publication was created, with its main objective being the introduction of the ES most relevant to the territory of Slovakia

The contents of this publication include a pilot assessment of the important ESs, with a total of 18 – 5 provisioning, 10 regulating/supporting, and 3 cultural ESs – selected for the territory of Slovakia. In the characterization of the ES, information from various available sources were used, in particular scientific reviews and articles. The specific ES assessment is based on available spatial and information datasets, all ESs have been assessed by a comparable methodological approach. The result of the assessment is a relative scale expressing the suitability of a territory for the provision of the given ES within the whole territory of Slovakia. This scale can be replaced in the future by specific biophysical units or monetary values based on specialized research or a value transfer method – this is the path which can be seen as the most perspective. The relationship between the ES and the main types of landscape and ecosystems and the importance of the particular ES in terms of nature and landscape protection are also assessed. The conclusion of this publication provides a summary ES comparison according to their basic groups, an overall assessment and proposals for further continuation of the ES assessment process in Slovakia.

The publication was prepared with the financial support of the MoE SR (Ministry of Environment of Slovak Republic), using the results of several scientific projects and available spatial and information datasets of different research institutions – SNC SR, Constantine the Philosopher University in Nitra and the Institute of Landscape Ecology of the SAS (Slovak Academy of Sciences). The ambition of the publication is to present a summary of available theoretical and methodological knowledge and to introduce the results of the first phase of a comprehensive ES assessment for the territory of Slovakia. The authors of the publication hope that they will have the opportunity to follow up on the presented results (also in the wider author's collective) in the next stage of the ES assessment, which should focus on the assessment of the ES demand, their real flow in the landscape, and last but not least on the economic (monetary) ES assessment in Slovakia. For this purpose, case studies from different territories, carried out in different scales and by different methodologies, could also prove useful, as these can also help to better assess the ES at the national level.

Constantine the Philosopher University in Nitra
Nitra, Slovakia
Peter Mederly

State Nature Conservancy of the Slovak Republic
Banská Bystrica, Slovakia
Ján Černecký

Authors:
Peter Mederly[1], Ján Černecký[1,2,3], Jana Špulerová[3], Zita Izakovičová[3], Martin Jančovič[1], Viktória Ďuricová[2,4], František Petrovič[1], Simona Gusejnov[1], Juraj Hreško[1], Dagmar Štefunková[3], Barbora Šatalová[3], Matej Močko[1], Gréta Vrbičanová[1], Dominika Kaisová[1], Martina Turanovičová[1], Tomáš Kováč[1] and Ivan Laco[1]

[1] Faculty of Natural Sciences, Constantine the Philosopher University in Nitra, Tr. A. Hlinku 1, 949 74 Nitra, Slovakia; [2] State Nature Conservancy of the Slovak Republic, Tajovského 28B, 974 01 Banská Bystrica, Slovakia; [3] Institute of Landscape Ecology of the Slovak Academy of Sciences, Štefánikova 3, P.O.Box 254, 814 99 Bratislava, Slovakia; [4] Faculty of Natural Sciences, Matej Bel University, Tajovského 40, 974 01 Banská Bystrica, Slovakia

Authors of Maps and Graphs:
Peter Mederly and Matej Močko

Authors of Photographs:
Ján Černecký, Dominika Kaisová, Dagmar Štefunková, Jana Špulerová, Ján Hreško, František Petrovič, Martin Jančovič, Stanislav David and Ľudmila Černecká

Author of Ecosystem Services Icons:
Martina Turanovičová

Acknowledgments

This book was created with the support of project *Monitoring of species and habitats of Community Interest according to the Habitat and Bird Directives* ITMS 310011P170 (implemented within Operational program Quality of Enviroment) and Slovak APVV grant agency project APVV-17-0377 "*Assessment of recent changes and trends in agricultural landscape of Slovakia.*"

Contents

Contributors

Ján Černecký Faculty of Natural Sciences, Constantine the Philosopher University in Nitra, Nitra, Slovakia

State Nature Conservancy of the Slovak Republic, Banská Bystrica, Slovakia

Institute of Landscape Ecology of the Slovak Academy of Sciences, Bratislava, Slovakia

Viktória Ďuricová State Nature Conservancy of the Slovak Republic, Banská Bystrica, Slovakia

Faculty of Natural Sciences, Matej Bel University, Banská Bystrica, Slovakia

Simona Gusejnov Faculty of Natural Sciences, Constantine the Philosopher University in Nitra, Nitra, Slovakia

Juraj Hreško Faculty of Natural Sciences, Constantine the Philosopher University in Nitra, Nitra, Slovakia

Zita Izakovičová Institute of Landscape Ecology of the Slovak Academy of Sciences, Bratislava, Slovakia

Martin Jančovič Faculty of Natural Sciences, Constantine the Philosopher University in Nitra, Nitra, Slovakia

Dominika Kaisová Faculty of Natural Sciences, Constantine the Philosopher University in Nitra, Nitra, Slovakia

Tomáš Kováč Faculty of Natural Sciences, Constantine the Philosopher University in Nitra, Nitra, Slovakia

Ivan Laco Faculty of Natural Sciences, Constantine the Philosopher University in Nitra, Nitra, Slovakia

Peter Mederly Faculty of Natural Sciences, Constantine the Philosopher University in Nitra, Nitra, Slovakia

Matej Močko Faculty of Natural Sciences, Constantine the Philosopher University in Nitra, Nitra, Slovakia

František Petrovič Faculty of Natural Sciences, Constantine the Philosopher University in Nitra, Nitra, Slovakia

Barbora Šatalová Institute of Landscape Ecology of the Slovak Academy of Sciences, Bratislava, Slovakia

Jana Špulerová Institute of Landscape Ecology of the Slovak Academy of Sciences, Bratislava, Slovakia

Dagmar Štefunková Institute of Landscape Ecology of the Slovak Academy of Sciences, Bratislava, Slovakia

Martina Turanovičová Faculty of Natural Sciences, Constantine the Philosopher University in Nitra, Nitra, Slovakia

Gréta Vrbičanová Faculty of Natural Sciences, Constantine the Philosopher University in Nitra, Nitra, Slovakia

Abbreviations

ARIES	Artificial Inteligence for Ecosystem Services
AUSLEM	Australian Land Erodibility Model
SEU	Evaluated Soil-Ecological Units
BT	Benefit Transfer
CBD	Convention on Biological Diversity
CICES	The Common International Classificastion of Ecosystem Services
CLC	Corine Land Cover
CV	Contingent Valuation
RS	Remote Sensing
EV	Economic Value
EAFRD	European Agricultural Fund for Rural Development
ES	Ecosystem Service
EST	Ecosystem Services Toolkit
ESTIMAP	Ecosystem Service Mapping Tool
EU	European Union
FAO	Food and Agriculture Organization of the United Nations
FAPAR	The Fraction of Absorbed Photosyntheticaly Active Radiation
GM	geomorphology mesoregion
GIS	Geographic Information System
GDP	Gross Domestic Product
HNV	High Natural Value
ESFT	Economic Set of Forest Types
PA	Protected Area
CLA	Characteristic Landscape Appearance
SPA	Special Protection Area
InVEST	Integrated Valuation of Ecosystem Services and Tradeoffs
IPBES	Intergovernmental Platform for Biodiversity and Ecosystem Services
CIMS	Comprehensive Information and Monitoring System
MaB	Man and Biosphere Programme
MEA	Millenium Ecosystem Assessment
MoE	Ministry of Environment of SR

NDVI	The Normalised Difference Vegetation Index
NLC	National Forest Centre
NPPC	National Agricultural and Food Centre
NNR	National Nature Reservation
NaLP	Nature and Landscape Protection
OP QE	Operational Programme Quality of Environment
OpenNESS	Operalisation of Natural Capital and Ecosystem Services
OZ	Civil association
HPZ	Hygienic Protection Zone
NR	Nature Reserve
RUSLE	Revised Universal Soil Loss Equation
SAS	Slovak Academy of Sciences
SEA	Slovak Environment Agency
SEEA	System of Environmental Economic Accounting
SEU	Soil-Ecological Units
SHMÚ	Slovak Hydrometeorological Institute
CLS	Current Landscape Structure
SAU	Slovak Agricultural University
SR	Slovak Republic
STU	Slovak Technical University
SWME	Slovak Water Management Enterprise
SGIDŠ	State Geological Institute of Dionýz Štúr
SNC SR	State Nature Conservancy of the Slovak Republic
TEEB	The Economics of Ecosystems and Biodivesity
TESSA	Toolkit for Ecosystem Service Site-based Assessment
PG	Permanent Grassland
TU	Technical University
UMB	University of Matej Bel
UNESCO	The United Nations Educational, Scientific and Cultural Organization
UKNEA	The UK National Ecosystem Assessment
UKF	Constantine the Philosopher University
USD	US Dollar
USLE	Universal Soil Loss Equation
USPED	Unit Stream Power-based Erosion Deposition
SAC	Special Area of Conservation
ILE	Institute of Landscape Ecology
TSES	Territorial System of Ecological Stability
SSCRI	Soil Science and Conservation Research Institute
WRI	Water Research Institute
WTA	Willingness to accept
WTP	Willingness to pay
Coll.	Collection of Laws
Env.	Environment

List of Figures

List of Tables

Part I
Ecosystem Services Basics

Chapter 1
Overview of the Ecosystem Services Concept

Peter Mederly, Ján Černecký, Simona Gusejnov, Viktória Ďuricová, and Gréta Vrbičanová

Abstract The chapter provides a concise introduction into the issue of ecosystem services (ES), its research, and implementation. It starts with the definition, primary model, and a brief history of this concept. The ES approach is increasingly reflected and applied in the scientific field, but it does not yet have the necessary support in the economic and decision-making areas. The second part is devoted to the ES classification systems, which are used for the research, followed by an overview of basic assessment methods. The most used classification of ES includes provisioning, regulating and supporting, and cultural ES. Research methods are usually divided into biophysical, sociocultural, and economic (monetary). The most frequently used and recommended methods are briefly described. The next part characterizes the level of ES research and implementation in the European Union, which is a leader in this field. Significant progress was achieved in most EU Member States within the Mapping and Assessment of Ecosystems and their Services (MAES) process. The last part of the chapter provides the ES political background and research outputs in Slovakia. The implementation rate of the ES concept in the Slovak Republic is one of the lowest in the whole of the EU – and this is a real challenge for the future.

P. Mederly (✉) · S. Gusejnov · G. Vrbičanová
Faculty of Natural Sciences, Constantine the Philosopher University in Nitra, Nitra, Slovakia
e-mail: pmederly@ukf.sk; simona.stasova@ukf.sk; greta.vrbicanova@ukf.sk

J. Černecký
Faculty of Natural Sciences, Constantine the Philosopher University in Nitra, Nitra, Slovakia

State Nature Conservancy of the Slovak Republic, Banská Bystrica, Slovakia

Institute of Landscape Ecology of the Slovak Academy of Sciences, Bratislava, Slovakia
e-mail: jan.cernecky@sopsr.sk;

V. Ďuricová
State Nature Conservancy of the Slovak Republic, Banská Bystrica, Slovakia

Faculty of Natural Sciences, Matej Bel University, Banská Bystrica, Slovakia
e-mail: viktoria.duricova@sopsr.sk

P. Mederly, J. Černecký (eds.), *A Catalogue of Ecosystem Services in Slovakia*,
https://doi.org/10.1007/978-3-030-46508-7_1

1.1 Ecosystem Services Basics

1.1.1 Introduction to Ecosystem Services

Although the concept of ecosystem services or utility properties and functions of nature began to show up in scientific literature, social, and political debates already in the late 1960s and early 1970s, key research and a broader discussion on this issue can only be dated back to the late 1990s. In particular, ecosystem functions and services depend on the quality and quantity of natural resources (such as soil, air, and water) and biodiversity – referred to overall as natural capital. Therefore, it is necessary to assess the ecosystem services (hereinafter referred to as ES) in relation to the functions, processes, and structure of related ecosystems, that is, to the quality of the environment of a particular territory and the value of its natural capital.

By using natural resources, including ecosystem services and with other activities, people directly or indirectly affect the natural environment and the quality of its components, in both temporal and spatial framework – including short-, medium-, and long-term local, regional, and global scales. The interaction of nature and man is the basis for the concept of ecosystem services – the ES is thus a concept between natural, social, and economic sciences.

The ES concept focuses on comprehensive research of ecosystems, their functions, and the assessment of benefits which ecosystems can provide for the society. It is constructed on an interdisciplinary basis. The ES concept seeks to ensure the protection and efficient use of ecosystems and their services so that all ESs are harmoniously used and that one ES is not developed further to the detriment of others. Several methods have been developed for ES assessment, including monetary and non-monetary, participative, and biophysical. An important part of the concept is the involvement of various groups engaged in ES assessment and management (Izakovičová et al. 2017).

The main idea of the ES concept is therefore the usefulness and benefits of nature for the society and human well-being. Ecosystem services can be very easily defined as contributions of ecosystems (living systems) to human well-being. These services are final (end) as they present outputs of ecosystems (whether natural, semi-natural, or largely altered by human activity) which directly affect human well-being. Their basic attribute is that they retain a link to the related ecosystem functions, processes, and ecosystem structure itself, which co-creates them (Haines-Young and Potschin 2013).

1.1.2 Cascade Model of ES

A clear formulation of the ES concept is provided by the so-called cascade model (Haines-Young and Potschin 2018), which clearly defines the sequence of notions of ecosystem structure and processes – ecosystem functions – ecosystem services – benefits from ecosystem services – service values (see Fig. 1.1).

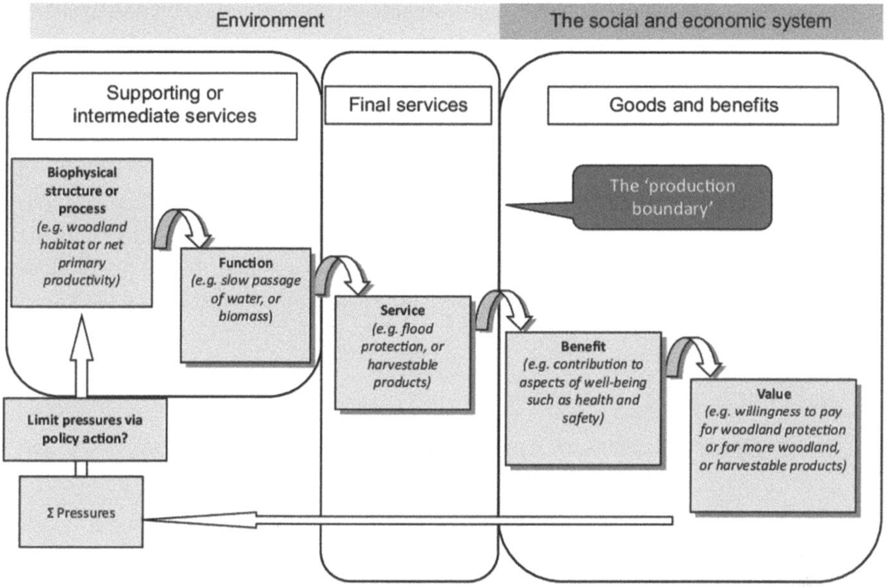

Fig. 1.1 Cascade model – from structure to functions, services, benefits, and values. (Source: Haines-Young and Potschin 2018)

The cascade model can be interpreted as follows:

- The ecosystems themselves, more precisely the geoecosystems, are the cornerstone and basic premise for the functioning of the ES. In the model, they are represented by a set of biophysical structures or processes which encompass the entire set of ecological components (e.g., matter, energy, and species), as well as key ecological processes (e.g., nutrient and energy cycles) taking place within the ecosystem. Obviously, only healthy ecosystems can provide a good quality ES – therefore, terms such as resilience, stability, and ecosystem integrity are accentuated.
- The next stage of the cascade is formed by ecosystem functions – these include the ecological components and processes that have the capacity to generate benefits used by people and thereby directly or indirectly support economic activities. According to Gómez-Baggethun et al. (2010), these represent a key link between ecology and economics.
- The central position of the cascade includes the ESs which, in a sense, represent the final outputs of the ecosystem – they are linked to ecosystem structures and processes, but at the same time, they are directly involved in generating benefits used by humans. Their existence is conditioned by the existence of demand and consumption of these services – without human use, they would not be considered ecosystem services.
- The final stage of the cascade consists of goods and benefits, representing the social and economic system. They are specific because they have a specific value

for humans – either monetary or non-monetary. The benefit can be understood as a concrete contribution of the ES to human well-being, with the value being its concrete valuation. It can be expressed differently, not only financially, because humans also attribute importance to the benefits based on moral, aesthetic, or spiritual values.

– With the use of ecosystem services as well as through the intermediary impacts on ecosystem functions and via other ways of influencing the landscape, humans put pressure on real geoecosystems, thereby causing adverse changes in their structure and functions and thus in further potential for their use. This feedback is shown in the model by an arrow which points away from the values back to the left side of the model.

1.1.3 A Brief History of Application of the ES Concept

The notion of ES was comprehensively explained for the first time in a publication Ehrlich and Ehrlich (1981) and has since been gradually applied, especially in scientific publications. Approximately since 2000, the establishment of the concept in the political agenda can be observed – for example through the so-called ecosystem approach, adopted in the year 2000 at the 5th Conference of the *Convention on Biological Diversity* in Nairobi, Kenya.

An ES summary vision and its basic classification, which has been used in the world literature, has been compiled by a large-scale project Millennium Ecosystem Assessment (MEA) in the period from 2001 to 2005 (project synthesis is presented in MEA 2005). The study of Costanza and Daly (1992), which estimated the average annual value of 17 selected ecosystem services at $33 trillion per year, which is approximately 1.5 times the global economy's GDP, has also been widely medialized. This value has been updated and refined to $125 trillion for 2011 (Costanza et al. 2014) – but with changes in landscape use and anthropogenic impacts since 1997, the ES value dropped by $20.2 trillion worldwide.

The economic dimension of the concept in 2010 was highlighted by the study The Economics of Ecosystems and Biodiversity (TEEB 2010) or by the wider TEEB initiative, which aims to *enhance the visibility of natural values*. Along with the increasing number of studies on monetary assessment of ecosystem services, the interest of decision-making and policymaking bodies has gradually begun to shift more towards the prospective creation of market-based instruments which could provide economic incentives for nature conservation.

At the global level, the ESs have also been established through the CBD. The Strategic Plan for Biodiversity 2011–2020 also includes the so-called *Aichi biodiversity targets*, two of which are particularly relevant for the ES (objectives 1 and 2, in more detail below). The establishment of an Intergovernmental Science-Policy Platform on Biodiversity and Ecosystem Services (IPBES) in 2012 has also helped to better integrate the ES concept into the policy agenda. IPBES creates a science and policy interface which enables scientific findings and analysis to be communicated towards the decision-making bodies and apply them also in the framework of international conventions. An example can be found in the Regional Assessment Report on Biodiversity and Ecosystem Services for Europe and Central Asia (IPBES 2018). Among the scientific and expert forums for research and promotion of the ES concept, it is appropriate to mention the Ecosystem Services Partnership (the largest international global network for ES research and application – available online: www.es-partnership.org) and the Natural Capital Project (a partnership between several universities and international organizations in the area of the development of ES assessment tools and their enforcement in decision-making – available online: www.naturalcapitalproject.org).

In addition to policy initiatives, the interest of the private sector in the ES concept has been growing in recent years. One such initiative is, for example, the Natural Capital Coalition, which brings together various stakeholders with a common vision to create a world where private companies protect and maintain the natural capital.

In their work, Costanza et al. (2017) describe the development of the ES concept in relative detail and very clearly at the same time. This article is highly recommended to those interested in this topic. They see the main progress in research and application of the ES concept in the following areas: the transition from definitions to classifications and ES assessment, the transition from integrated modelling to public participation and communication, and the development of institutions and innovations in the societal governance.

The fact that the ES concept is increasingly reflected and applied in the scientific field is also proven by the analysis of scientific and research publications on this issue in the Scopus scientific database (available online: www.scopus.com). By the end of 2000, approximately 150 articles and publications focused on the ES were published, compared to April 2017 when Costanza et al. mentioned approximately 17,000 titles containing the term ecosystem services in the title, abstract, or keywords. By the end of 2018, this number increased to 23,880 documents, and in the first 3 months of 2019, we saw an additional 1310 contributions (see Fig. 1.2).

This development can also be documented in the *Ecosystem Services* journal, which is the *flagship* of ES research. The journal was founded in 2012 by two prominent members of the scientific community of this topic (Rudolf de Groot and Leon Braat). In the first three-year period, there were 405 articles published in the journal, and in April 2019, this number increased to 880 articles altogether. In the first years, the journal published an average of 5–6 articles a month; in the year 2016, it was already 10–12 articles; and now it is more than 15 articles a month. However, it is only a fraction of what is published in all the world's research and scientific periodicals.

Despite the above, it is clear that the ES concept does not yet have the necessary support, especially in the economic field, nor in the area of important decision-making. At the end of the assessment study, Costanza et al. (2017) state the following:

> In particular, it points to the weakness of the mainstream economic approaches to valuation, growth, and development. The substantial contributions of ecosystem services to the sustainable wellbeing of humans and the rest of nature should be at the core of the fundamental change needed in economic theory and practice if we are to achieve a societal transformation to a sustainable and desirable future.

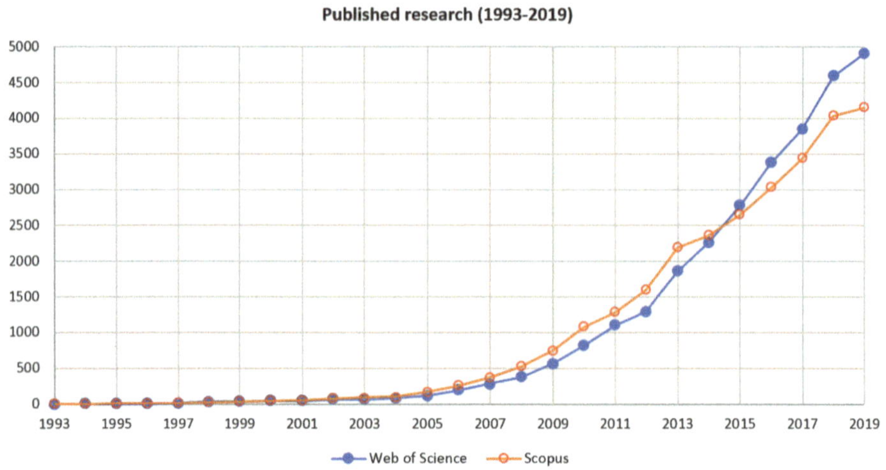

Fig. 1.2 Development of publications focusing on ecosystem services, included in the Web of Science and Scopus database

1.2 Ecosystem Services Classification

The basic classification of ecosystem services includes *provisioning services, regulating and supporting services, and cultural services*. There are a number of approaches to their more detailed classification, with the best-known being classification performed within the Millennium Ecosystem Assessment (MEA 2005) project, The Economics of Ecosystems and Biodiversity project classification (2010) and the Common International Classification of Ecosystem Services – CICES (Haines-Young and Potschin 2018). MEA 2005 is the basic classification used globally, especially before 2010, and CICES is the most detailed classification used for ES hierarchical classification and assessment.

Basic classification of ES according to CICES (Haines-Young and Potschin 2018) is the following:

- Provisioning services – this includes all material products and goods from ecosystems, providing nutrition, materials, and energy, especially *biomass for nutrition, drinking water and water for other non-drinking purposes, utility biomass, abiotic materials, and energy sources.*
- Regulating and supporting services – this includes the benefits from ecosystem functions regulating natural processes, as well as ecosystem functions and processes relevant to the healthy state of ecosystems and the provision of other services, in particular:

  Regulating services – regulation of waste, toxic substances, and other pollutants; regulation and mediation of flows (mass, liquid, and gaseous); regulation and protection of life cycles and habitats; and regulation and control of pests and diseases

Supporting services – in particular physical, chemical, and biological conditions: soil formation and composition, water cycle and water conditions, atmospheric composition and climate regulation, and other supporting processes

- Cultural services – this includes non-material benefits from ecosystems and biotic features of the landscape: *physical and experiential interactions, intellectual and representative interactions, spiritual and emblematic interactions, and other cultural outcomes.*

A more detailed description of the individual ES is provided in the main assessment part of the publication. A comparison of the ES basic classification systems is given in Table 1.1.

Table 1.1 Main classification systems of ecosystem services

ES group	Costanza et al. 1997	Category pursuant to MEA 2005	Category pursuant to TEEB 2010	Category pursuant to CICES – Haines-Young and Potschin 2018
Provisioning services	Food production	Food	Food	Biomass – nutrition Freshwater and sea plants and animals for nutrition
	Water supply	Fresh water	Water	Ground and surface water for drinking
				Ground and surface water for non-drinking purposes
	Raw materials	Fibre, timber	Raw materials	Utility biomass – timber and other fibres
	Genetic resources	Genetic resources	Genetic resources	Genetic sources of biotic origin
		Biochemicals and natural medicines	Medicinal resources	Genetic material for biochemical and pharmaceutical processes
	x	Ornamental resources	Ornamental resources	Materials of biotic origin (ornamental resources)
	x	x	x	Biomass – Sources of energy of plant and animal origin
	x	x	x	Abiotic sources of energy

(continued)

Table 1.1 (continued)

ES group	Costanza et al. 1997	Category pursuant to MEA 2005	Category pursuant to TEEB 2010	Category pursuant to CICES – Haines-Young and Potschin 2018
Regulating and supporting services	Gas regulation	Air quality regulation	Air purification	Regulation of gaseous and air flows
	Waste treatment	Water purification and waste treatment	Waste treatment (esp. water purification)	Regulation of waste, toxic substances, and other pollutants
	Disturbance regulation (storm protection and flood control)	Natural hazard regulation	Disturbance prevention or mediation	Regulation of air and liquid flows
	Water regulation (e.g., natural irrigation and drought prevention)	Water regulation	Regulation of water flows	Regulation of liquid flows
	Erosion control and sediment retention	Erosion regulation	Erosion prevention	Regulation (mediation) of mass flows
	Climate regulation	Climate regulation	Climate regulation	Atmospheric composition and global climate regulation
	Soil formation	Soil formation (supporting service)	Soil fertility maintenance	Support of soil formation and composition
	Pollination	Pollination	Pollination	Lifecycle maintenance (including pollination)
	Refuges (nursery, migration habitats)	Biodiversity	Lifecycle maintenance (esp. nursery) Gene pool protection	Life cycle and habitats maintenance, gene pool protection
	Biological control	Regulation of pests and diseases	Biological control	Support of pest and disease control
	Nutrient cycling	Nutrient cycling and photosynthesis, primary production	x	x

(continued)

Table 1.1 (continued)

ES group	Costanza et al. 1997	Category pursuant to MEA 2005	Category pursuant to TEEB 2010	Category pursuant to CICES – Haines-Young and Potschin 2018
Cultural services	Recreation (incl. ecotourism and outdoor activities)	Recreation and ecotourism	Recreation and ecotourism	Physical and experiential interactions (recreation and ecotourism)
	Cultural (incl. aesthetic, artistic, spiritual, education, and science)	Aesthetic values	Aesthetic information	Experiential interactions
		Cultural diversity	Inspiration for culture, art, and design	Representative interactions (promotion, art)
		Spiritual and religious values	Spiritual experience	Spiritual and/or emblematic interactions (cultural heritage)
		Knowledge systems and educational values	Information for cognitive development	Intellectual interactions (willingness to protect nature, moral aspects)

Source: Costanza et al. (2017), modified

1.3 Basic Assessment Methods of Ecosystems Services

ES assessment is a complex and multidisciplinary issue, and when dealing with this issue, it is not appropriate to remain at the level of scientific methods. For example, Gómez-Baggethun et al. (2010) report that assessing ecosystems and their services should not be seen as a goal, but as a pragmatic tool pointed to the assessment of the true contribution of nature to human well-being and its incorporation into economic theory and practical decision-making. Jacobs et al. (2014) state that the ultimate objective of ES assessment is to contribute to a more sustainable and fair use of natural resources. Accordingly, Daily (2000) has proposed human well-being as a unit for ES assessment, with the aim of ES assessment being the improvement of the well-being of the whole society while respecting the principles of sustainability (ensuring the needs of the present generation without compromising the needs of future generations).

Majority of ES experts agree that a number of methods are appropriate for ES assessment – but in principle, it is possible to summarize them into three basic groups according to the main principle of assessment and provision of results –bio-physical methods, sociocultural (non-monetary) methods, and economic (monetary) methods. In addition, there are integrated methods which use multiple approaches and often combine multiple methods. From the point of view of the purpose of the assessment used, Costanza et al. (2017) recognize methods aimed at raising public awareness and interest, economic accounting, specific policy analysis, spatial development and land use planning, payments for ES, cost accounting, and general asset management.

Neugarten et al. (2018) provide an overview of ecosystem assessment tools based mainly on biophysical assessment and modelling (Table 1.2). At the same time, they created a *decision tree* for the selection of methods (Fig. 1.3).

The following text provides an overview and a brief description of the most frequently used and recommended methods of ES assessment – more specifically, the methods are presented in the characteristics of individual ES in the main part of this publication.

1.3.1 Biophysical/Natural Science Methods

Ecological (biophysical) assessment is usually the first step in ES assessment. It focuses in particular on assessment of the condition and functioning of ecosystems and their characteristics, from which the social and economic values are consequently derived. According to de Groot et al. (in Jacobs et al. 2014), the ecological value includes the ecosystem health with ecological indicators such as diversity or integrity. In the System of Environmental-Economic Accounting (SEEA; European Commission 2014), the value in biophysical units represents the quantification of the flow of assessed services where the ESs are expressed as material and energy flows.

In order to express the ES value, measurable indicators are most commonly used, and in justified cases, substitute indicators (proxy-indicators) can be used. Mathematical and biophysical models (hydrological, climatic, erosion, production, etc.) are used to express the state, functions, and processes in ecosystems as well as the ES potential. Specific mapping methods are also often used – for instance, based on geographic information systems – and allow for spatial rendering of the value or ES provision and their components (e.g., ES matrix method – Burkhard et al. 2009, 2014).

According to Gomez-Baggethun and De Groot et al. (2010), the main biophysical method includes the following:

- Ecological footprint – describes the spatial extent of the biologically productive area which the society uses for its consumption – inputs and outputs (similar are, e.g., carbon or water footprint)
- Land cover flow analysis – used to monitor changes in natural capital quality and soil multifunctionality
- Material flow analysis – monitors environmental inputs and outputs within the socio-economic system metabolism

Table 1.2 Overview of ecosystem services assessment tools

Tool name and acronym	Internet source	Citation
Tools written *step by step*		
Ecosystem Services Toolkit – EST	publications.gc.ca/site/eng/9.829253/publication.html	Value of Nature to Canadians Study Taskforce (2017)
Protected Areas Benefits Assessment Tool – PA-BAT	wwf.panda.org/our_work/biodiversity/protected_areas/arguments_for_protection/	Dudley and Stolton (2008); Ivanić et al. (2017)
Toolkit for Ecosystem Service Site-based Assessment v.2.0 – TESSA	tessa.tools	Peh et al. (2017)
Computer model-based tools		
ARtificial Intelligence for Ecosystem Services – ARIES	aries.integratedmodelling.org	Villa et al. (2009)
Co$ting Nature v.3 – C$N	www.policysupport.org/costingnature	Mulligan (2015)
Integrated Valuation of Ecosystem Services and Tradeoffs 3.4.2 – InVEST	www.naturalcapitalproject.org/invest/	Sharp et al. (2018)
Multiscale Integrated Models of Ecosystem Services – MIMES	www.afordablefutures.com	Boumans et al. (2015)
Social Values for Ecosystem Services – SolVES	solves.cr.usgs.gov	Sherrouse et al. (2011)
WaterWorld v.2 – WW	www.policysupport.org/waterworld	Mulligan (2013)

Source: Neugarten et al. (2018)

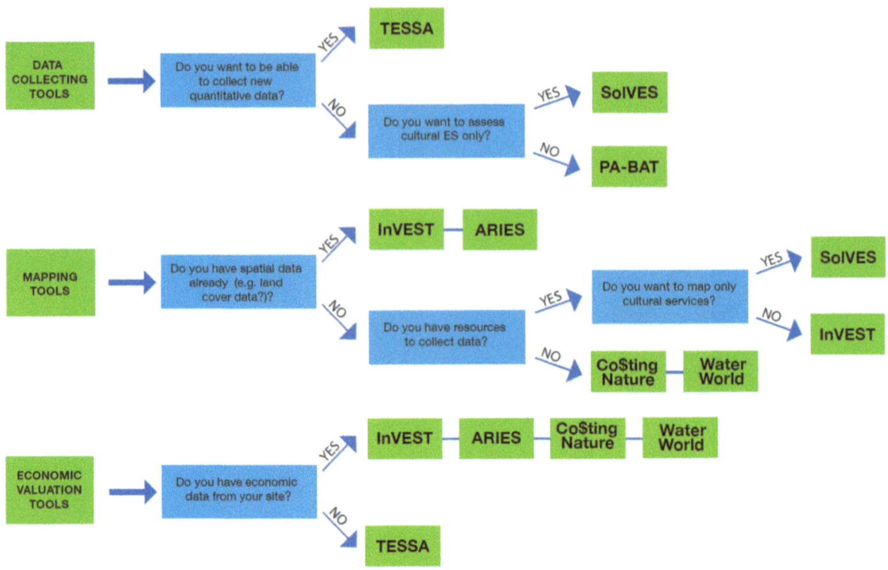

Fig. 1.3 Decision process (tree) for ES tool selection. (Source: Neugarten et al. 2018)

- Life cycle analysis – monitors the process of a certain activity or production cycle from its creation to its completion (liquidation, termination)
- Energy/exergy methods – aim to quantify the amount of energy that needs to be introduced during the performance of a given (e.g., economic) process

The best known (predominantly biophysical) models used for ES assessment include the following:

- InVEST: a set of spatial biophysical models for quantifying and assessing ES benefits created at Stanford University, suitable particularly for local and regional level (available online: www.naturalcapitalproject.org). Some models are also designed to describe the ES economic value.
- ESTIMAP: the spatial model used mainly on the continental scale but with several applications also on the national level. It enables the assessment of the impact of different land use change scenarios on ES provision. Eight analytical models focusing on different regulating ES (e.g., pollination, air quality regulation) are operational at EU level, but the module for assessment of recreational ES is most commonly used (Zulian et al. 2013, 2018).
- QuickScan: a comprehensive software spatial-statistical tool designed for participative decision-making by representatives of various stakeholder groups and subjects with the participation of experts (Verweij et al. 2016). The model can be used for different purposes and in different scales (available online: www.quickscan.pro).

1.3.2 Sociocultural Methods

There are a number of issues *outside the domain* of natural sciences which are related to the assessment of ecosystem services, including those from the social, cultural, and historical context of this issue. It is therefore logical to use *inclusive* assessment with the involvement of stakeholder group representatives (this term will also be used in other parts of the publication) and the related use of other than classical natural science methods.

Sociocultural assessment is understood either as a subset or as a synonym of the so-called non-monetary ES assessment which is focused on the *importance, preferences, needs, or requirements which people express in relation to nature* (de Groot et al. 2010; Chan et al. 2016; Castro et al. 2014). The number of studies using these methods for ES assessment is still growing, and so the sociocultural methods are becoming an accepted part of the ES concept, although they still do not have fully established methodological background (Gómez-Baggethun et al. 2014).

Sociocultural methods are mostly based on *qualitative data* – especially on value estimates or the importance of individual ES, they express the social preferences of people and population groups with respect to the ES. These are the so-called *deliberative methods* which use, for example, the expression of relative significance instead of monetary or economic values. They are often based on collective and interactive procedures – for example workshops, meetings, structured interviews, or questionnaire methods. So, it is not so much about determining the exact value (for example, the suitability of the territory for the provision of the given ES), rather than attaining approval, or agreement on a particular assessment or solution.

Sociocultural assessment includes a wide range of methods, the most commonly used of which are the following (according to Santos-Martín et al. 2017):

- Preference assessment – consultation method for analysing the perception, recognising, and assessment of the demand or use of the ES.
- Time use methods – determining respondents' willingness to devote time to changing ES quality or quantity
- Photo-elicitation survey – exploring the value of a particular place in terms of ES provision based on respondents' perceptions and feelings
- Narrative methods – methods using description or specific story to express ecosystem/landscape value from an ES perspective
- Participatory mapping – ES assessment with participation and application of knowledge of various stakeholders of the society
- Scenario planning – creating possible future scenarios and assessing their relationship with ES utilization (usually with participatory methods)
- Deliberative methods – assessment and decision-making (including ES issue) through an open discussion of stakeholder representatives

1.3.3 Economic/Monetary Methods

Considering that most ESs are public goods (not directly part of the market), their economic value is usually not adequately reflected in market-based processes, and thus they are threatened by overuse or deterioration. One of the main goals of economic assessment is to avoid such a scenario by better reflecting the economic value of the ES into the decision-making processes.

In this context, it is especially the issue of assessment of the so-called externalities (such as the related effects and costs of ES use, which are not directly included in the ES price) and their incorporation into economic accounting and decision-making processes – this process is the domain of *environmental economics*. To this end, economists use mainly the concept of *total economic value*, which is composed of both use and non-use values. To capture these values, economics uses a variety of methods – primary methods or value transfer methods. For primary methods, *direct market methods* (in particular market prices and interactions) are used – if such information is not available, then parallel or hypothetical markets based on *preference surveys* are used. If no such data is available or a survey cannot be conducted directly in the research area, then the information obtained in other research is used, i.e. the mentioned *transfer of values*.

Overview of used economic (especially monetary) methods of ES assessment:

- *Direct valuation methods:* in particular market price, avoided damage method, prevention cost, restoration cost, production function, spared government spending, and others – consist of a direct ES financial valuation
- *Revealed preference methods:* travel costs, hedonic pricing, opportunity costs – an estimate of the ES values through similar real functions or services in the landscape
- *Stated preference methods:* contingent valuation, choice experiments – an estimate of the ES value through the preferences (statements) of the respondents
- *Benefit/value transfer methods:* ES valuation in model territory based on research from existing primary assessment studies from other territories or political contexts

In conclusion, it needs to be pointed out that the attitudes on the ES monetary assessment vary. Although most scientists recognize its need (especially as a tool to raise awareness or to compare the cost of different alternatives to improve ES provision), some authors argue the usefulness of economic assessment. For example, according to Spangenberg and Settele (2010), the ES monetary assessment fails to capture the ES value in a broader sense, ignoring their social and ecological qualities perceived by ES beneficiaries at different levels. Norgaard (2000) states that current monetary assessment methods only help us see ES values from an unsustainable economic point of view and not from the *desirable* sustainable economic model. The ethical dimension of nature services assessment is also frequently discussed (e.g., Chan et al. 2016; Jax et al. 2013). Overall, there resonates a need for the economic assessment to be broadened into a wider ES assessment context with its main role as a supporting tool for moving towards a sustainable society.

1.3.4 Integrated Assessment of ES

Given the complexity of assessing the ES issues and the value pluralism associated with it (de Groot et al. 2002; Gómez-Baggethun and de Groot 2010; Jax et al. 2013), there is a consensus in the scientific community concerning the need to link different ES assessment methods and the development of the so-called *integrated assessment methods*. Significant progress has been made in the area of integrated ES assessment in recent years – in particular through scientific projects aimed at transferring research results to management and decision-making practice (OpenNESS and ESMERALDA projects – see below in Sect. 1.4).

An overview of various methods which can be used for the ES assessment and the interconnections between them is shown in Fig. 1.4 – it is a summary of the methods used in the OpenNESS project according to Barton and Harrison (2017). Obviously, interpreting the results achieved using a variety of different methods is not easy – the integrated methods should formalize and facilitate the process. For example, the framework for integrated assessment is also mentioned by Gómez-Baggethun and Barton (2013) – according to them, it is necessary to define the purpose of the assessment and the policy context, the degree of accuracy required, spatial resolution, and geographic scale – and only then select the appropriate methods.

In general, integrated methods are used for the *overall assessment* of the final benefits of the ESs for human well-being or quality of life. They also help with the decision on priorities for the use of individual ESs, which are expressed in different units and different methods. For this purpose, the following are used, for instance:

- Multi-criteria decision analysis (MCDA) – a participatory tool used to link ecological, sociocultural, and economic contexts through an assessment and discussion framework involving various stakeholder groups (a specific policy framework), using modelling.
- Bayesian belief networks (BBN) – probabilistic models (charts) for decision-making in different probability conditions. They allow the gradual creation of a model decision network and assessing their likely consequences.
- State and transition models (STM) – expert modelling of the probable changes in the state of ecosystems, their properties, and their functions due to various decisions. They can be linked to spatial geographic information system (GIS) models.
- Scenario development – defining several possible directions for further development of a certain territory, based on verified assumptions about substantial trends and drivers. It is important to involve stakeholders in this process.
- Deliberative valuation – it is not a method but rather an assessment framework, based on a combination of multiple methods and techniques, involving research-

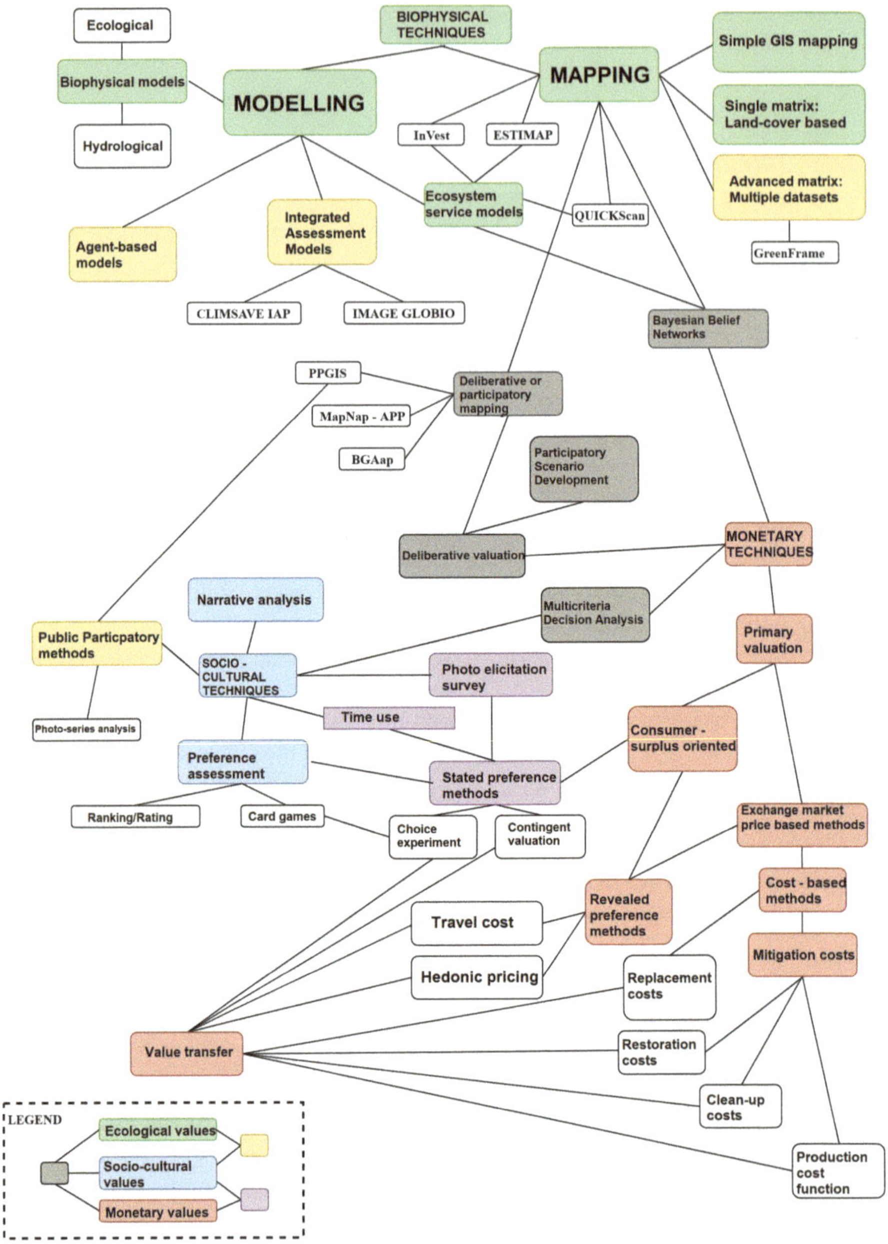

Fig. 1.4 Chart of methods used in ES assessment and relations between them. Groups of methods are coloured; examples of specific methods are given on a white background. (Source: Barton and Harrison (eds.) 2017, modified)

ers and representatives of different stakeholder groups. The result is achieved by mutual discussion and open dialogue, preferably by the consensus of a majority.

Several of these methods, or procedures, not only are *integration* but also can be described as combined – they also use the techniques of biophysical, sociocultural, and partly economic assessment.

Since the issue of integrated assessment is very complex, it cannot be summarized in a limited extent. Those interested in this area of research can find more information for further study in the work by Barton and Harrison (2017).

1.4 Process of Ecosystem Services Assessment in the European Union

1.4.1 Policy Context of ES Assessment

The EU became one of the leaders of the research and implementation of the ecosystem services concept. In particular, after 2010, the EU has adopted several important documents in the field of natural resources protection and biodiversity promotion – from the 1998 strategy through 2001 and 2006 action plans to the current EU biodiversity strategy 2020 adopted in 2011. The introduction of this strategy emphasizes the importance of biodiversity as part of natural capital in terms of ES provision and the overall standard of living (quality of life) of people. The strategy aims to reverse the loss of biodiversity and accelerate the EU's transition to a resource-efficient *green* economy.

The vision of EU biodiversity policy by 2050 is the protection, valuation, and adequate restoration of biodiversity and ecosystem services (natural capital) it provides. The main reason is the intrinsic value of biodiversity and its fundamental contribution to the standard of living and economic prosperity. The main goal by 2020 is to stop the loss of biodiversity and ES degradation within the EU and restore them to the fullest extent possible while increasing the EU's contribution to preventing global biodiversity loss.

The EU 2020 biodiversity strategy consists of 6 targets and 20 actions focused on halting biodiversity loss and the degradation of ecosystem services. ESs were included in target no. 2 Maintaining and enhancing ecosystems and their services, which specifies the following:

> By 2020, ecosystems and their services are maintained and enhanced by establishing green infrastructure and restoring at least 15% of degraded ecosystems (European Commission 2011).

Special emphasis on ES has been transferred into action no. 5:

> Member States, with the assistance of the Commission, will map and assess the state of ecosystems and their services in their national territory by 2014, assess the economic value of such services, and promote the integration of these values into accounting and reporting systems at EU and national level by 2020 (European Commission 2011).

In order to support this goal, the European Commission has initiated the creation of an expert group on MAES (available online: www.biodiversity.europa.eu/mae). Within this group, much progress has been achieved in most EU Member States in the area of ES assessment. More support was provided by the EU-funded international scientific projects, in particular, OpenNESS (available online: www.openness-project.eu), aimed at operationalizing the concept of natural capital and ES; OPERAs (available online: www.operas-project.eu), focusing on how these concepts can be transferred from academia to practice; and ESMERALDA (available online: www.esmeralda-project.eu), which builds on both previous projects, with the goal to create a flexible methodology for ES assessment at European level as well as regional or local level.

A valuable output of the OpenNESS project includes 27 model studies at local and regional levels in 13 European and 4 non-European countries (for more information on outputs – Wijnja et al. 2016; Dick et al. 2018). An interesting tool could also be found in the Oppla platform (available online: www.oppla.eu), which is an open *marketplace* of knowledge about ES, natural capital, and nature-based solutions where experts from various fields – science, research and practice, public and private sectors, individuals, and small and large organizations – may find answers to related questions (Izakovičová et al. 2017).

1.4.2 National ES Assessments in Europe

National ES assessments in European countries are one of the main outcomes of the biodiversity protection strategy commitments and the functioning of the MAES working group. As part of the ESMERALDA project, a so-called *MAES barometer* has been prepared and is assessed. The barometer maps progress across individual countries (Fig. 1.5). According to this assessment, some countries have already achieved full implementation (not only the ecosystem and ES assessment but also

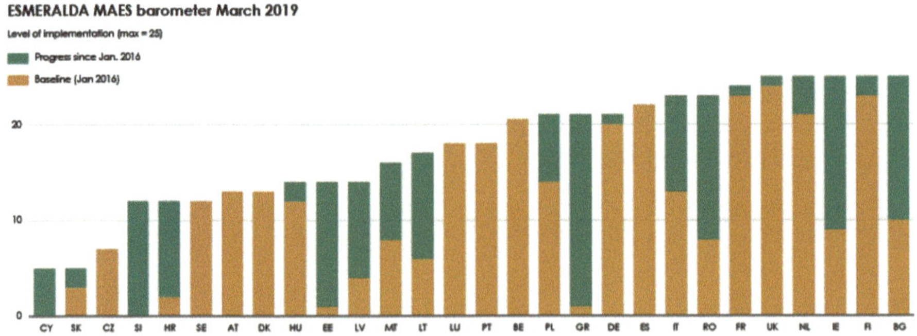

Fig. 1.5 ESMERALDA MAES barometer: EU Member States' progress in the assessment and implementation of the ES concept in the period 01/2016–03/2019. (Source: Biodiversity Information System for Europe (https://biodiversity.europa.eu/maes))

their integration in national policies) – the countries include the United Kingdom, the Netherlands, Ireland, Finland, and Bulgaria. Other countries are significantly approaching this objective (Italy, Romania, France). For the period since 2015, Greece, Ireland, Bulgaria, Romania, Estonia, and Slovenia have made the largest progress. The overall level of implementation (valid on March 2019) was assessed at 70% – unfortunately, Slovakia only reaches 20% and is at the very end of the ranking together with Cyprus.

The unflattering position of the SR is a consequence of the halting of the MAES process practically in the very beginning and the absence of financial resources to ensure a national ES assessment.

Table 1.3 provides an overview of national ES assessments with available information according to literature analysis and work by Schröter et al. (2016). In addition to the specified countries, several countries do not have national assessments available and published in English, or in progress of preparation (Bulgaria, Hungary, Italy, France, Greece). In addition to EU countries, Norway, Russia, and Israel are included in the table, for which background studies were available.

Table 1.3 Overview of national assessments of ES and the number of assessed ES by main groups

Country	ES total	Provisioning ES	Regulating and supporting ES	Cultural ES	Citation
Czech Republic (CZ)	18	7	5 / 4	2	Frélichová et al. (2014), Vačkář et al. (2018)
Denmark (DK)	11	3	1 / 2	5	Turner et al. (2014)
Finland (FI)	28	10	8 / 4	6	Jäppinen and Heliölä (2015)
Flanders (BE)	16	5	6 / 4	1	Stevens et al. (2015)
Netherlands (NL)	19	5	5 / 5	4	CBS (2015), PBL Netherlands (2019)
Ireland (IE)	28	9	5 / 6	8	Parker et al. (2016)
Lithuania (LT)	31	14	6 / 5	6	Depellegrin et al. (2016)
Luxembourg (LU)	13	4	4 / 4	1	Becerra-Jurado et al. (2016)
Germany (DE)	18	5	5 / 5	3	Rabe et al. (2016), Albert et al. (2016), Grunewald et al. (2016)
Romania (RO)	12	4	3 / 2	3	NEPA. (2017)
Russia (RU)	19	4	6 / 4	5	Bukvareva et al. (2017)
Spain (SP)	22	7	4 / 4	7	Santos-Martín et al. (2016)
Great Britain (UK)	26	12	4 / 5	5	UK NEA (2011)
Portugal (PT)	6	3	0 / 3	4	Schröter et al. (2016)
Norway (NO)	26	7	5 / 5	9	Schröter et al. (2016)
Israel (IS)	3*	0	0 / 3	1	Lotan et al. (2018)
Italy (IT)	5*	0	2 / 2	1	Giarratano et al. (2018)

The numbers in the cells indicate the number of assessed ES
*Assessments of only some ES available

The analysis of the studies shows some generalizations which can also be used for the process of preparing the ES national assessment in Slovakia. Here are the main facts:

- The number of ES for assessment in individual countries varies significantly but is on average 15–20 ES. The lowest number (3–6 ES) is reported by IS, IT, and PT; by contrast, the largest number (26–28 ES) is reported by NO, UK, FI, and IE.
- The emphasis on ES representation by main groups varies – some countries have over-represented provisioning ES (FI, LT, GB), other cultural ES (DK, IR, NO, SP). Regulating and supporting ES are significantly represented in almost all countries.
- Ecosystem maps were used as an important basis for the ES assessment for most countries. Some countries (LT, RU) used simpler land use maps or Corine Land Cover maps.
- Most countries use other indicators for the assessment of the ES – the natural environment properties database is standard, and it is further used for the selection of indicators, the creation of maps in the GIS, and the possible use of models. The most sophisticated indicator system is used by FI, LU, IE, UK, BE, and NL.
- ES assessment methods vary significantly across countries. Simple methods include mainly the use of the so-called assessment matrix (ES matrix – Burkhard et al. 2009, 2014) – this was used as the main method in, for example, national studies of LI and RU.
- More complex procedures in the form of ES mapping, ES indicators, and statistical data evaluation were, for example, presented in studies of BE, NL, UK, RO, and SP.
- Biophysical models have been used for the ES part in different countries – DK, FI, DE, IE, IT, and LU.
- The economic valuation of the ES in the form of the benefit transfer method was used by CZ, IT, UK, FI, and SP.
- Most of the studies focus on the current status and trends related to ES value, but some also offer future development scenarios (UK, PT, SP).
- Most of the studies address not only the ES capacity issues but also the demand and current ES flow issues and compare them in different ways. The most common include statistical evaluations of relationships between these categories for administrative units – regions (e.g., DK, DE).

1.5 Ecosystem Services Assessment in the Slovak Republic

1.5.1 Policy Process of ES Assessment

As is clear from the previous text, the implementation rate of the ES concept in the SR is one of the lowest in the whole of EU. However, this is mainly due to political factors, not a lack of expertise or necessary data. Unfortunately, in the previous period, there was not enough *political will* to ensure the assessment process, even though this process is required by the approved documents. The assessment of the current state of application of the ES concept in Slovakia at the political level (planning and decision-making at the national, regional, and local level) is part of the study by Bezák et al. (2017), which, in addition to analysing the current situation, also provides the basis for better implementation of the ES concept.

In 2012, following the adoption of the European strategy and tasks defined by the strategy, the SR prepared the Updated National Strategy for Biodiversity Protection for 2012–2020. The strategy was adopted by the Government Decree no. 12/2014 (MoE SR 2014). The aim is to create a policy framework to halt the loss of biodiversity and to accelerate the transition of the SR as an EU member country to the *green* economy, which uses natural genetic resources in accordance with the Europe 2020 Strategy. The *key objective* of the strategy is to halt the loss of biodiversity and the degradation of ecosystems and their services in the SR by 2020, to restore biodiversity and ecosystems to an appropriate extent, and to increase our contribution to preventing global biodiversity loss.

The vision set by the SR in this document is as follows:

> Natural Capital of the SR – biodiversity, ES and related goods are sufficiently protected by 2050, regularly assessed, wisely used and, where appropriate, restored due to their intrinsic values and for their significant contribution to the welfare and economic prosperity of the Slovak Republic. Adopted measures and policies at the national level prevent the adverse changes which the loss of natural capital would cause.

The strategy includes nine objectives, which are largely based on European objectives. Each of them focuses on a specific issue, with Objectives 1–3 being established for the protection and restoration of biodiversity and related ES. In particular, Objective 3 is important from an ES perspective:

> Ensure the maintenance and strengthening of ecosystems and their services by 2020 through the establishment of green infrastructure and the restoration of at least 15% of degraded ecosystems.

The following measures are important in particular for the identification, assessment, and subsequent protection of ecosystems and their services:

- Improve knowledge of ecosystems and services provided through mapping and assessing the status of ecosystems and their services in the SR
- Prepare a system of assessment and economic valuation of ES and goods and propose a comprehensive system of payments for ES use, taking into account existing systems and mechanisms

However, the achievement of this ambitious objective and the measures outlined above is unrealistic within the given timeframe by 2020 – therefore, the objectives will have to be revised or deadlines moved.

Following the national strategy for biodiversity, the ES concept was also transformed into the Environmental Policy Strategy of SR 2030 entitled Greener Slovakia (approved by the Slovak Government in February 2019). One of the measures is to assess and sustainably use the ES. In relation to ES, the environmental strategy states the following:

> By 2030, all ESs will be taken into account equally and shall be implemented within the national accounting system. The ESs will be assessed and quantified and taken into account when considering investments and policies as well as in environmental impact assessments. The establishment of a comprehensive ES assessment system and sustainable use of ESs will be supported and the possibilities for monetization will be considered. Payments for ESs will create sufficient incentives to maintain them.

These measures will also be developed in the Nature and Landscape Protection Framework by 2030.

In 2014, an expert working group MAES was established under the Ministry of the Environment, focusing on the achievement of Target 2 of the EU Biodiversity Strategy, i.e. mapping and assessing ecosystems and services provided by them. The group met more regularly in the period from 2014 to 2016 and met again in 2018. The group consists mainly of representatives of various ministerial professional organizations and institutions, academia, and local governments. Experts from the SNC SR were also part of the expert group, and they started the preparation of several activities and documents necessary for the assessment of the ES at the national level. An initial ecosystem map of Slovakia was prepared (Černecký et al. 2020), using data from various sectors (mainly from nature protection, agriculture, and forestry). In 2019, the verification process of the map commenced by botanists directly in the field (25 SNC SR employees) – in the first year, about 10% of the Slovak territory should be verified.

In the period from 2017 to 2018, Slovakia was represented by MoE SR in the international project ESMERALDA, funded by the EU Framework Program for research and innovation – Horizon 2020. Representatives of all EU Member States as well as some associated countries participated in the project. The project established a flexible methodology for mapping and assessing ecosystems and services provided by these ecosystems on a pan-European, national, and regional level. One of the outputs was the so-called MAES Explorer, a publicly available online tool to help implement EU Biodiversity Strategy Target 2 (available online: http://www.maes-explorer.eu/). Another tool provided was the so-called Methods Explorer, which provides a clear structured database of ES mapping and assessment methods.

Other activities related to the ES concept worth mentioning include in particular the systematic monitoring of habitats and species of community interest (66 habitat types and 196 species), which is an important database necessary for the assessment of many ES aspects. As part of the monitoring since 2013 under the professional guidance of SNC SR, comprehensive monitoring is conducted on more than 10,000 permanent monitoring sites. It is the largest field data collection in the history of Slovakia, which has so far involved more than 400 experts. The first stage of monitoring consisted of field collection, processing, and evaluation of data on the status of individual habitats and species of European importance. The results of this project and further information are available in publications by Šefferová Stanová and Galvánková (2015) and Janák et al. (2015). At the same time, the Comprehensive Information and Monitoring System (CIMS; available online: www.biomonitoring.sk)

was established, which aggregates the occurrence data on habitats and species in Slovakia provided by experts and the general public. In the current programming period, the Monitoring II project is approved, and SNC SR is preparing two larger projects with nationwide coverage. The first project is focused on management measures in non-forest habitats and the second on nature-based forest management in protected areas.

Another possibility and opportunity for improving the state of knowledge and implementation of management measures in the field of biodiversity protection and the ES is the Operational Programme Quality of Environment (OP QE), which is a programme document of the SR for drawing assistance from the EU Structural Funds and the Cohesion Fund in the 2014–2020 programme period. In terms of ES assessment, it is important to develop projects under priority axis 1 Sustainable use of natural resources through the development of environmental infrastructure, especially in investment priority 2.2 Biodiversity and soil protection and restoration and ES support, including NATURA 2000 and green infrastructure network. This priority offers opportunities to finance activities and measures for the conservation and improvement of habitats or ecosystems and thus directly supports the provision of ES in Slovakia. However, the support from the operational programme is limited by the duration of the programme period, and therefore, it is necessary to introduce systematic financing for support, restoration, and conservation of habitats in Slovakia by the MoE SR. The next step should be to involve small owners – local stakeholders – in the restoration of biodiversity and support them financially, for example through Envirofond.

The above-mentioned processes implemented by MoE SR (especially SNC SR) are a basic prerequisite for an adequate ES assessment. Notwithstanding, much more accurate and diverse data would be needed for a comprehensive ES assessment, but these are not currently being collected and are not supposed to be collected in the near future. Essentially, basic data sources lack quality and quantity, because data is often outdated, inaccurate, or incomplete. Despite the unfavourable situation, SNC SR is actively preparing a monograph in this area, which will, upon completion, present a national ES assessment from the perspective of an ecosystem approach based on the above-mentioned data sources.

1.5.2 Expert Level of ES Assessment

Although the ES concept is not as politically well established in Slovakia as in other European countries, its application has been gradually increasing in recent years, especially in the *expert field,* for example, in the valuation of functions and services of nature in protected areas, assessment of forest functions, agricultural soils assessment, assessment of historical agricultural landscape structures, and others.

The issue of ES research and assessment in the SR is currently investigated as part of *the research tasks and scientific projects of various workplaces*, with partial results and case studies being published (active workplaces in this area include mainly the Institute of Landscape Ecology of SAS, National Forest Centre, and National Agricultural and Food Centre). This also applies to research conducted by Slovak universities, which is fragmented into research projects and tasks of individual entities (especially Comenius University in Bratislava, Slovak University of Technology in Bratislava, Constantine the Philosopher University in Nitra, Slovak University of Agriculture in Nitra, Technical University in Zvolen, Matej Bel University in Banská Bystrica). Most of the existing spatial and database materials and partly the research capacities are concentrated in two research organizations within the MoE (Slovak Environmental Agency and State Nature Conservation). Research coordination and joint projects are rare, so the exchange of experience and presentation at various professional and scientific events is more implemented.

PhD research at some universities and research organizations is also focused on the education and preparation of ES specialists (11 dissertations with the ES topic were prepared in the SR in the period from 2014 to 2018, of which 5 were prepared at Constantine the Philosopher University in Nitra and 3 at the Slovak University of Agriculture in Nitra). The ES issues are also addressed in the final thesis of students of the above-mentioned universities (in the same period, approximately 75 bachelor and master theses focused on the ES assessment in general or in a particular territory – most at the Slovak University of Agriculture in Nitra, Matej Bel University in Banská Bystrica, Technical University in Zvolen, and Slovak University of Technology in Bratislava).

Some of the *first research publications* comprehensively assessing the non-production functions of forest ecosystems and vegetation in Slovakia generally include Papánek (1978), Midriak et al. (1981), and Jurko (1990), with Eliáš (1983, 2010) also focusing on this issue for a long period of time.

The ES concept is relatively best elaborated for forestry and nature conservation areas. The topic of forest ES and their assessment is mainly addressed by the researchers of the National Forest Centre and Technical University in Zvolen (e.g., Čaboun et al. 2008; Kovalčík and Tutka 2008; Čaboun et al. 2010, 2014; Konôpka 2010, 2012; Sarvašová and Šálka 2012; Šálka and Dobšinská 2013; Sarvašová et al. 2014; Štěrbová 2017; Šálka et al. 2017). In terms of theory, this topic was elaborated by, for example, Vološčuk (2013); also, the publication Schneider et al. (2016) could be useful for the Slovak studies. From the ES assessments of protected areas, it is possible to mention several publications – assessment of the Tatranský národný park (Fűzyová et al. 2009; Brezovská and Holécy 2009; Švajda 2009; Fleischer et al. 2017), National Park (NP) Slovenský raj (Getzner 2009), NP Veľká Fatra (Považan et al. 2014a), Nízkotatranský NP (Špulerová et al. 2016) and NP Muránska Planina (Považan et al. 2015). More generally, the ES of protected areas was mainly the topic of Považan et al. (2014b).

In the field of soil science and agriculture, attention was initially given to soil production functions (the concept is summarized, for example, in Džatko 2002). Especially after 2000, researchers began to put more attention on the complex of non-production soil functions (e.g., Hronec et al. 2005; Bujnovský et al. 2009; Tutka et al. 2009; Bujnovský 2011). Among the more recent studies, we can mention especially the articles of Vilček (2011, 2014), Vilček and Koco (2018), Kanianska (2014), Kanianska et al. (2016), Makovníková et al. (2016, 2017), and

Kizeková et al. (2016, 2018). The leader in this field is NPPC – Soil Science and Conservation Research Institute and Matej Bel University in Banská Bystrica.

Hydrological ES of Slovakia research is being developed within the Water Research Institute (Bujnovský 2018) and Slovak Agricultural University in Nitra (Juršík et al. 2017).

The assessment of the ES of *historical structures of the agricultural landscape* is mainly addressed by the Institute of Landscape Ecology of SAS (e.g., Špulerová 2006; Špulerová et al. 2014, 2017, 2018; Lieskovský et al. 2015). The assessment of the functions and services of vegetation in the residential environment is addressed by, for example, Supuka et al. 1991, 2000, Reháčková and Pauditšová 2006, and Turanovičová and Rózová 2017.

Participatory mapping and socio-economic assessment of ES have been addressed by, foe example, Bezák and Bezáková (2014), Kluvánková-Oravská and Chobotová (2010), Kluvánková-Oravská et al. (2013), and Kluvánková and Brnkaľáková (2017).

Of the more extensive ES assessment studies in specific model territories, it is worth to mention the case study of the OpenNESS EU project, which was conducted in the period from 2013 to 2016 at two institutions (Institute of Landscape Ecology SAS and Regioplan Nitra). In addition to analysing the current state of application of the ES concept in Slovakia (Bezák et al. 2017), the study also focused on the elaboration and direct implementation of several methods of ES assessment on the example of the model area of Trnava and its functional urban area (Mederly et al. 2017). Based on all project outputs, a proposal for appropriate landscape and spatial planning procedures has been developed, particularly with regard to the integration of the ES concept into the planning and decision-making process (Izakovičová et al. 2017).

1.5.3 Background for the ES Catalogue

Finally, as a starting point for the ES assessment in Slovakia, it is appropriate to quote the conclusions from the article by Izakovičová et al. 2017, which summarizes the results of the OpenNESS case study in the Trnava model territory:

> The ES concept is relatively unknown in Slovak terms, as evidenced by the results of the conducted research. Given the prevailing sectoral approach to the planning process in Slovakia and the poor application of integrated policies or strategies, the implementation of the ES concept is quite limited. The ES are not reflected in national strategic documents or laws which would be binding for the local implementation of spatial policies. The public interests represented by the ES are suppressed by local, mostly individual preferences. The ES concept, which represents an integrated approach to landscape assessment with a focus on participatory methods, has a great potential to streamline spatial planning in Slovakia. When considering the effective implementation of the ES concept, the following will be necessary:
>
> – implement the ES concept into the environmental policy and legislation, i.e. to change the legislation of spatial planning and nature and landscape protection, and subsequently modify landscape documentation methodologies;

- implement the integrated principles and participative methods of the ES concept in spatial-planning processes and reflect the ES concept in sectoral plans while harmonizing the objectives of sectoral policies;
- develop a national ES strategy in Slovakia and develop a national, regional and local ES assessment methodology;
- set up stakeholders at different spatial levels to support the implementation of the ES concept and overcome gaps, by the top-down approach – from national strategies to local implementation;
- focus on the mandatory incorporation of the ES concept, in particular in local strategies, which are mandatory as part of EU funding applications (e.g., PESD);
- ensure effective education, training and dissemination.

References

Albert, C., Bonn, A., Burkhard, B., Daube, S., Dietrich, K., Engels, B., Frommer, J., Götzl, M., Grêt-Regamey, A., Job-Hoben, B., Koellner, T., Marzelli, S., Moning, C., Müller, F., Rabe, S. E., Ring, I., Schwaiger, E., Schweppe-Kraft, B., & Wüstemann, H. (2016). Towards a national set of ecosystem service indicators: Insights from Germany. *Ecological Indicators, 61*, 38–48.

Barton, D. N., & Harrison, P. (Eds.). (2017). *Integrated valuation of ecosystem services. Guidelines and experiences*. EU FP7 OpenNESS Project Deliverable, European Commission FP7. https://oppla.eu/sites/default/files/uploads/openness-d3344integratedassessmentvaluationofesfinal2.pdf. Accessed 4 Apr 2019.

Becerra-Jurado, G., Philipsen, C., & Kleeschulte, S. (2016). *Mapping and assessing ecosystems and their services in Luxembourg – Assessment results*. Luxembourg: Le Gouvernment du Grand -Duché de Luxemburg. https://doi.org/10.13140/rg.2.1.4924.5841.

Bezák, P., & Bezáková, M. (2014). Landscape capacity for ecosystem services provision based on expert knowledge and public perception (case study from the North West Slovakia). *Ekologia Bratislava, 33*(4), 344–353. https://doi.org/10.2478/eko-2014-0031.

Bezák, P., Mederly, P., Izakovičová, Z., Špulerová, J., & Schleyer, C. H. (2017). Divergence and conflicts in landscape planning across spatial scales in Slovakia: Opportunity for an ecosystem services-based approach? *International Journal of Biodiversity Science, Ecosystem Services & Management, 13*(2), 119–135. https://doi.org/10.1080/21513732.2017.1305992.

Boumans, R., Roman, J., Altman, I., & Kaufman, L. (2015). The Multiscale Integrated Model of Ecosystem Services (MIMES): Simulating the interactions of coupled human and natural systems. *Ecosystem Services, 12*, 30–41. https://doi.org/10.1016/j.ecoser.2015.01.004.

Brezovská, K., & Holécy, J. (2009). Ocenenie rekreačnej funkcie lesov Vysokých Tatier metódou cestovných nákladov. *Acta Facultatis Forestalis Zvolen, LI*(1), 151–162. https://lf.tuzvo.sk/sites/default/files/aff_09_suppl_1.pdf. Accessed 4 Apr 2019.

Bujnovský, R. (2011). Hodnota poľnohospodárskej pôdy a jej ekologických funkcií. *Životné prostredie, 45*(1), 35–37. http://147.213.211.222/sites/default/files/2011_1_035_037_bujnovsky.pdf

Bujnovský, R. (2018). Estimation of benefits from the actual use of inland water ecosystem services in the Slovak Republic. *Ekológia, 37*(3), 201–218. https://doi.org/10.2478/eko-2018-0017.

Bujnovský, R., Balkovič, J., Barančíková, G., Makovníková, J., & Vilček, J. (2009). *Hodnotenie a oceňovanie ekologických funkcií poľnohospodárskych pôd Slovenska*. Slovakia: VÚPOP.

Bukvareva, T., et al. (2017). Supplied, demanded and consumed ecosystem services: Prospects for national assessment in Russia. *Ecological Indicators, 78*(1), 351–360. https://doi.org/10.1016/j.ecolind.2017.03.034.

Burkhard, B., Kroll, F., Muller, F., & Windhorst, W. (2009). Landscapes' capacities to provide ecosystem services – A concept for land-cover based assessments. *Landscape Online, 15*, 1–22. https://doi.org/10.3097/LO.200915.
Burkhard, B., Kandziora, M., Hou, Y., & Müller, F. (2014). Ecosystem service potentials, flows and demands – Concepts for spatial localisation, indication and quantification. *Landscape Online, 34*(1), 1–32. https://doi.org/10.3097/LO.201434.
Čaboun, V., Moravčík, M., & Tutka, J. (2008). *Výskum, klasifikácia a uplatňovanie funkcií lesov v krajine. Lesy a lesníctvo – riziká, výzvy, riešenia*. Slovakia: National Forest Centre.
Čaboun, V., Tutka, J., Moravčík, M., Kovalčík, M., Sarvašová, Z., Schwarz, M., & Zemko, M. (2010). *Uplatňovanie funkcií lesa v krajine*. Slovakia: National Forest Centre.
Čaboun, V., Kovalčík, M., & Sarvašová, Z. (2014). Concept of the integrative aspects of the forest ecosystem services with case study on recreation services assessment in Slovakia. *Ecological Processes, 15*(3), 1–6. https://doi.org/10.1186/s13717-014-0015-8.
Castro, A. J., Verburg, P. H., Martín-López, B., Garcia-Llorente, M., Cabello, J., Vaughn, C. C., & López, E. (2014). Ecosystem service trade-offs from supply to social demand: A landscape-scale spatial analysis. *Landscape and Urban Planning, 132*, 102–110. https://doi.org/10.1016/j.landurbplan.2014.08.009.
CBS et al. (2015). *Ecosystem goods and services in the Netherlands*. https://www.clo.nl/en/indicators/en1572-ecosystem-goods-and-services-in-the-netherlands. Accessed 2 Mar 2019.
Černecký, J., Gajdoš, P., Špulerová, J., Halada, Ľ., Mederly, P., Ulrych, L., Ďuricová, V., Švajda, J., Černecká, Ľ., Andráš, P., & Rybanič, R. (2020). Ecosystems in Slovakia. *Journal of Maps, 16*, 28–35. https://doi.org/10.1080/17445647.2019.1689858.
Chan, K., Balvanera, P., Benessaiah, K., et al. (2016). Why protect nature? Rethinking values and the environment. *PNAS, 113*, 1462–1465. https://doi.org/10.1073/pnas.152502113.
Costanza, R., & Daly, H. (1992). Natural capital and sustainable development. *Conservation Biology, 6*, 37–46.
Costanza, R., dArge, R., de Groot, R., Farber, S., Grasso, M., Hannon, B., Limburg, K., Naeem, S., Oneill, R. V., Paruelo, J., Raskin, R. G., Sutton, P., & van den Belt, M. (1997). The value of the world's ecosystem services and natural capital. *Nature, 387*(6630), 253–260.
Costanza, R., De Groot, R., Sutton, P. C., Van Der Ploeg, S., Anderson, S., Kubiszewski, I., Farber, S., & Turner, R. K. (2014). Changes in the global value of ecosystem services. *Global Environmental Change, 26*, 152–158.
Costanza, R., De Groot, R., Braat, L., Kubiszewski, I., Fioramonti, L., Sutton, P., Farber, S., & Grasso, M. (2017). Twenty years of ecosystem services: How far have we come and how far do we still need to go? *Ecosystem Services, 28*, 1–16. https://doi.org/10.1016/j.ecoser.2017.09.008.
Daily, G. (2000). The value of nature and the nature of value. *Science, 289*(5478), 395–396.
De Groot, R. S., Wilson, M. A., & Boumans, R. M. (2002). A typology for the classification, description and valuation of ecosystem functions, goods and services. *Ecological Economics, 41*(3), 393–408.
De Groot, R. S., Alkemade, R., Braat, L., Hein, L., & Willemen, L. (2010). Challenges in integrating the concept of ecosystem services and values in landscape planning, management and decision making. *Ecological Complexity, 7*, 260–272.
Depellegrin, D., Pereira, P., Misiunė, I., & Egarter-Vigl, L. (2016). Mapping ecosystem services potential in Lithuania. *International Journal of Sustainable Development & World Ecology, 23*(5), 441–455. https://doi.org/10.1080/13504509.2016.1146176.
Dick, J., Turkelboom, F., Woods, H., et al. (2018). Stakeholders' perspectives on the operationalisation of the ecosystem service concept: Results from 27 case studies. *Ecosystem Services, 29*, 552–565. https://doi.org/10.1016/j.ecoser.2017.09.015.
Dudley, N., & Stolton, S. (2008). *The protected areas benefits assessment tool: A methodology*. Gland: WWF International. Available at: d2ouvy59p0dg6k.cloudfront.net/ downloads/pa_bat_web_1355739158.pdf. Accessed 13 Apr 2019.

Džatko, M. (2002). *Hodnotenie produkčného potenciálu poľnohospodárskych pôd a pôdno-ekologických regiónov Slovenska*. Slovakia: Research Institute of Soil Science and Soil Protection.

Ehrlich, P. R., & Ehrlich, A. H. (1981). *Extinction: The causes and consequences of the disappearance of species*. New York: Random House.

Eliáš, P. (1983). Ecological and social functions of vegetation. *Ekológia, 2*, 93–104.

Eliáš, P. (2010). Od funkcií vegetácie k ekosystémovým službám. *Životné prostredie, 44*(2), 59–64.

European Commission (2011). *European Commission COM/2011/0244*. Communication from the commission to the European parliament, the Council, the Economic and Social Sommittee and the Sommittee of the regions.

European Commission. (2014). *System of environmental-economic accounting 2012 experimental ecosystem accounting*. New York: European Commission Organisation for Economic Co-operation.

Fleischer, P., Pichler, V., Fleischer, P., Holko, L., et al. (2017). Forest ecosystem services affected by natural disturbances, climate and land-use changes in the Tatra Mountains. *Climate Research, 73*(1–2), 57–71. https://doi.org/10.3354/cr01461.

Frélichová, J., Vačkář, D., Pártl, A., Loučková, B., Harmáčková, Z. V., & Lorencová, E. (2014). Integrated assessment of ecosystem services in the Czech Republic. *Ecosystem Services, 8*, 110–117. https://doi.org/10.1016/j.ecoser.2014.03.001.

Fűzyová, Ľ., Lániková, D., & Novorolský, M. (2009). Economic valuation of Tatras national park and regional environmental policy. *Polish Journal of Environmental Studies, 18*(5), 811–818.

Getzner, M. (2009). *Economic and cultural values related to protected areas. Part A: Valuation of ecosystem services in Tatra (PL) and Slovensky Raj (SK) national parks*. Final report. http://www.minzp.sk/files/sekcia-ochranyprirodyakrajiny/ps-543/final-report-part-b-pl-sk_getzner_201003.pdf. Accessed 04 April 2019.

Giarratano, M. C., et al. (2018). *Comitato Capitale Naturale*. Roma: Secondo Rapporto sullo Stato del Capitale Naturale in Italia. https://www.minambiente.it/comunicati/il-secondo-rapporto-sullo-stato-del-capitale-naturale-italia. Accessed 6 Apr 2019.

Gómez-Baggethun, E., & Barton, D. (2013). Classifying and valuing ecosystem services for urban planning. *Ecological Economics, 86*, 235–245. ISSN 0921-8009.

Gómez-Baggethun, E., & De Groot, R. (2010). Chapter 5: Natural capital and ecosystem services: The ecological foundation of human society. In R. M. Harrison & R. E. Hester (Eds.), *Ecosystem services* (pp. 105–121). https://doi.org/10.1039/9781849731058-00105.

Gómez-Baggethun, E., De Groot, R., Lomas, P. L., & Montes, C. (2010). The history of ecosystem services in economic theory and practice: From early notions to markets and payment schemes. *Ecological Economics, 69*(6), 1209–1218. https://doi.org/10.1016/j.ecolecon.2009.11.007.

Gómez-Baggethun, E., et al. (2014). *State-of-the-art report on integrated valuation of ecosystem services*. EU FP7 OpenNESS project deliverable 4.1., European Commission FP7. http://www.openness-project.eu/sites/default/files/Deliverable%204%201_Integrated-Valuation-Of-Ecosystem-Services.pdf. Accessed 6 Apr 2019.

Grunewald, K., Herold, H., Marzelli, S., Meinell, G., Richter, B., Sybre, R. U., & Walz, U. (2016). Concept of national indicators for ecosystem services in Germany further development, types of classes and sheet of indicators. *In Naturschutz und Landschaftsplanung, 48*, 141–152.

Haines-Young, R., & Potschin, M. (2013). *CICES V4.3-Revised report prepared following consultation on CICES Version 4, August–December 2012*. Nottingham: Centre for Environmental Management, University of Nottingham. https://cices.eu/content/uploads/sites/8/2018/01/Guidance-V51-01012018.pdf. Accessed 6 Apr 2019.

Haines-Young, R., & Potschin, R. M. (2018). *Common International Classification of Ecosystem Services (CICES) V5.1 and guidance on the application of the revised structure*. Nottingham: Barton in Fabis. https://cices.eu/content/uploads/sites/8/2018/01/Guidance-V51-01012018.pdf. Accessed 2 Apr 2019.

Hronec, O., Bedrna, Z., & Vilček, J. (2005). *Environmentálna pedológia*. Slovakia: Slovak University of Agriculture in Nitra.

IPBES. (2018). *Summary for policymakers of the regional assessment report on biodiversity and ecosystem services for Europe and Central Asia of the Intergovernmental Science-Policy Platform on Biodiversity and Ecosystem Services*. https://www.ipbes.net/assessment-reports/eca. Accessed 6 Apr 2019.

Ivanić, K.-Z., Štefan, A., Porej, D., & Stolton, S. (2017). Using a participatory assessment of ecosystem services in the Dinaric Arc of Europe to support protected area management. *Parks, 23*, 61–74. https://doi.org/10.2305/IUCN.CH.2017.PARKS-23-1K-ZI.en.

Izakovičová, Z., Bezák, P., Mederly, P., & Špulerová, J. (2017). Uplatňovanie konceptu ekosystémových služieb v plánovacej a riadiacej praxi v Slovenskej republike – výsledky projektu OpenNESS na prípadovej štúdii Trnava. *Životné prostredie, 51*(4), 198–204.

Jacobs, S., Dendoncker, N., & Keune, H. (2014). *Ecosystem services: Global issues, local practices*. Boston: Elsevier.

Janák, M., Černecký, J., & Saxa, A. (Eds.). (2015). *Monitoring of animal species of Community interest in the Slovak Republic, Results and assessment in the period of 2013–2015*. Banská Bystrica: State nature conservancy of the Slovak Republic.

Jäppinen, J. P., & Heliölä, J. (Eds.). (2015). *Towards a sustainable and genuinely green economy. The value and social significance of ecosystem services in Finland (TEEB for Finland). Synthesis and roadmap*. Helsinki: The Finnish Ministry of Environment, Helsinki.

Jax, K., Barton, D. N., Chan, K. M. A., De Groot, R., Doyle, U., Eser, U., Görg, C. H., Gómez-Baggethun, E., Griewald, J., Haber, W., Haines-Young, R., Heink, U., Jahn, T., Joosten, H., Kerschbaumer, L., Korn, H., Luck, G. W., Matzdorf, B., & Wichmann, S. (2013). Ecosystem services and ethics. *Ecological Economics, 93*, 260–268. https://doi.org/10.1016/j.ecolecon.2013.06.008.

Jurík, Ľ., Halászová, K., & Sedmáková, M. (2017). Ekosystémové služby zdrojov vody pre krajinu. *Životné prostredie, 51*(4), 221–226.

Jurko, A. (1990). *Ekologické a socio-ekonomické hodnotenie vegetácie*. Bratislava: Píroda.

Kanianska, R. (2014). *Ekosystémové služby a ich oceňovanie. In Stav a trendy integrovaného manažmentu životného prostredia*. Slovakia, Zvolen: Technical University of Zvolen.

Kanianska, R., Jaďuďová, J., Makovníková, J., Kizeková, M., & Tomaškin, J. (2016). *Ekosystémové služby*. Banská Bystrica: Belianum.

Kizeková, M., Čunderlík, J., Dugátová, Z., Makovníková, J., Kanianska, R., Jaďuďová, J., Jančová, Ľ., & Pálka, B. (2016). *Agroekosystémové služby a súčasný stav trávnych porastov v Slovenskej republike*. Slovakia/Banská Bystrica: NPPC VÚTPHP Banská Bystrica, NPPC SSCRI Bratislava, UMB Banská Bystrica.

Kizeková, M., Čunderlík, J., Dugátová, Z., Jančová, Ľ., Kanianska, R., & Makovníková, J. (2018). *Klasifikácia ekosystémových služieb trávnych porastov a ich hodnotenie*. http://www.agroporadenstvo.sk/index.php?start&t=rastlinna-vyroba-rastlinna-vyroba&t2=&article=1118. Accessed 4 Apr 2019.

Kluvánková, T., & Brnkaľáková, S. (2017). Globálne problémy a lokálne riešenia v udržateľnom manažmente ekosystémových služieb a krajiny. *Životné prostredie, 51*(4), 195–197.

Kluvánková-Oravská, T., & Chobotová, V. (2010). Inštitúcie a ekosystémové služby v demokratickej a trhovej spoločnosti. *Životné prostredie, 44*(2), 84–88.

Kluvánková-Oravská, T., Ditmarová, Ľ., & Blaženec, M. (2013). *Ecosystem services: Interdisciplinary research challenges*. Paper presented at ecosystem services: Adaptive EU policies for future EU regions, Slovak Academy of Sciences, Bratislava, 30 September–1 October 2013.

Konôpka, J. (2010). Od reprodukcie k multifunkčnému využívaniu lesných ekosystémov. *Lesnícky časopis – Forestry Journal, 56*(1), 81–92.

Konôpka, J. (2012). Manažment hydrických funkcií lesov. *Lesnícky časopis – Forestry Journal, 58*(2), 129–135.

Kovalčík, M., & Tutka, J. (2008). *Hodnotenie rekreačnej funkcie lesov SR preferenčnými metódami – výsledky testovacej štúdie*. Paper presented at Aktuálne otázky ekonomiky LH SR, National Forest Centre, Zvolen, 2008.

Lieskovský, J., Bezák, P., Špulerová, J., Lieskovský, T., Koleda, P., Dobrovodská, M., Buergi, M., & Gimmi, U. (2015). The abandonment of traditional agricultural landscape in Slovakia – Analysis of extent and driving forces. *Journal of Rural Studies, 37*, 75–84. https://doi.org/10.1016/j.jrurstud.2014.12.007.

Lotan, A., Kost, R., Mandelik, Y., Peled, Y., Chakuki, D., Shamir, S. Z., & Ram, Y. (2018). National scale mapping of ecosystem services in Israel – Genetic resources, pollination and cultural services. *One Ecosystem, 3*, e25494. https://doi.org/10.3897/oneeco.3.e25494.

Makovníková, J., Kobza, J., Pálka, B., Mališ, J., Kanianska, R., & Kizeková, M. (2016). An approach to mapping the potential of cultural agroecosystem services. *Soil and Water Research, 11*(1), 44–52. https://doi.org/10.17221/109/2015-SWR.

Makovníková, J., Pálka, B., Širáň, M., Kanianska, R., Kizeková, M., & Jaďuďová, J. (2017). *Modelovanie a hodnotenie agroekosystémových služieb*. Banská Bystrica: Matej Bel University.

Mederly, P., Bezák, P., Lieskovský, J., Halabuk, A., Izakovičová, Z., & Dobrucká, A. (2017). Vybrané metódy hodnotenia ekosystémových služieb – projekt Open- NESS a prípadová štúdia Trnava. *Životné Prostredie, 51*, 205–212.

Midriak, R., et al. (1981). *Diferencované obhospodarovanie lesa podľa integrovaných funkcií* (Lesnícke štúdie 31). Zvolen: VÚLH & Bratislava: Príroda.

Millennium Ecosystem Assessment. (2005). *Ecosystems and human well-being: Wetlands and water synthesis*. Washington, DC: World Resources Institute.

MoE SR. (2014). *Aktualizovaná národná stratégia ochrany biodiverzity do roku 2020*. http://www.minzp.sk/sekcie/temy-oblasti/ochrana-prirody-krajiny/medzinarodne-dohovory/dohovor-bio-diverzite/aktualizovana-narodna-strategia-ochrany-biodiverzity-do-roku-2020/. Accessed 6 Apr 2019.

Mulligan, M. (2013). WaterWorld: A self-parameterising, physically based model for application in data-poor but problem-rich environments globally. *Hydrology Research, 44*, 748. https://doi.org/10.2166/nh.2012.217.

Mulligan, M. (2015) Trading off agriculture with nature's other benefits, spatially In Zolin, C. A and Rodrigues, R., de, A. R. (eds) Impact of climate change on water resources in agriculture. CRC Press. ISBN 9781498706148.

NEPA. (2017). *Assessment of ecosystems and ecosystem Services in Romania*. Romania: NEPA, NINA, ROSA, WWF Romania.

Neugarten, R. A., Langhammer, P. F., Osipova, E., Bagstad, K. J., Bhagabati, N., Butchart, S. H. M., Dudley, N., Elliott, V., Gerber, L. R., Gutierrez Arrellano, C., Ivanić, K.-Z., Kettunen, M., Mandle, L., Merriman, J. C., Mulligan, M., Peh, K. S.-H., Raudsepp-Hearne, C., Semmens, D. J., Stolton, S., & Willcock, S. (2018). *Tools for measuring, modelling, and valuing ecosystem services: Guidance for key biodiversity areas, natural world heritage sites and protected areas*. Gland: IUCN.

Norgaard, R. B. (2000). Ecological economics. *BioScience, 50*(4), 291. https://doi.org/10.1641/0006-3568(2000)050[0291:EE]2.3.CO;2.

Papánek, F. (1978). *Teória a prax funkčne integrovaného lesného hospodárstva*. Bratislava: Príroda.

Parker, N., Naumann, E.-K., Medcalf, K., Haines-Young, R., Potschin, M., Kretsch, C., Parker, J., & Burkhard, B. (2016). *Irish wildlife manuals, national ecosystem and ecosystem service mapping pilot for a suite of prioritised services – Appendices*. Ireland: National Parks and Wildlife Service, Department of Arts, Heritage, Regional, Rural and Gaeltacht Affairs.

Peh, K. S.-H., Balmford, A. P., Bradbury, R. B., Brown, C., Butchart, S. H. M., Hughes, F. M. R., MacDonald, M. A., Stattersfield, A. J., Thomas, D. H. L., Trevelyan, R. J., Walpole, M., & Merriman, J. C. (2017). *Toolkit for Ecosystem Service Site-based Assessment (TESSA), version 2.0*. Cambridge.

Považan, R., Getzner, M., & Švajda, J. (2014a). Value of ecosystem services in mountain national park. Case study of Veľká Fatra National Park (Slovakia). *Polish Journal of Environmental Studies, 23*(5), 1699–1710.

Považan, R., Getzner, M., & Kadlečík, J. (2014b). Hodnotenie ekosystémových služieb v chránených územiach Karpát so zameraním na Slovensko – Príručka pre rýchle hodnotenie. *Quaestiones rerum naturalium, 1*(2), 7–44.

Považan, R., Getzner, M., & Švajda, J. (2015). On the valuation of ecosystem services in Muránska Planina National Park (Slovakia). *Eco. Mont. Journal on Protected Mountain Areas Research and Management, 7*(2), 61–69. https://doi.org/10.1553/eco.mont-7-2s61.

Rabe, S.-E., Koellner, T., Marzelli, S., & Schumacher, P. (2016). Gret-Regamey, A. (2016). National ecosystem services mapping at multiple scales – The German exemplar. *Ecological Indicators, 70*, 357–372. https://doi.org/10.1016/j.ecolind.2016.05.043.

Reháčková, T., & Pauditšová, E. (2006). *Vegetácia v urbánnom prostredí*. Bratislava: Cicero.

Šálka, J., & Dobšinská, Z. (2013). *Policy analysis for assuring forest ecosystem externalities*. Zvolen: Technical University in Zvolen.

Šálka, J., Dobšinská, Z., & Štěrbová, M. (2017). *Analýza verejnej politiky na zabezpečenie ekosystémových služieb lesa*. Zvolen: Technical University in Zvolen.

Santos-Martín, F., García Llorente, M., Quintas-Soriano, C., Zorrilla-Miras, P., Martín-López, B., Loureiro, M., Benayas, J., & Montes, M. (2016). *Spanish National Ecosystem Assessment: Socio-economic valuation of ecosystem services in Spain.Synthesis of the key findings*. Madrid: Biodiversity Foundation of the Spanish Ministry of Agriculture, Food and Environment.

Santos-Martín, F., Kelemen, E., García-Llorente, M., Jacobs, S., Oteros-Rozas, E., Barton, D. N., Palomo, P., Hevia, V., & Martín-López, B. (2017). Socio-cultural valuation approaches. In B. Burkhard & J. Maes (Eds.), *Mapping ecosystem services* (pp. 104–114). Sofia: Pensoft Publishers.

Sarvašová, Z., & Šálka, J. (2012). *Integrácia úžitkov verejnoprospešných funkcií do trhového mechanizmu – teórie a zahraničné skúsenosti*. Zvolen: Technical University in Zvolen.

Sarvašová, Z., Kovalčík, M., Dobšinská, Z., Šálka, J., & Jarský, V. (2014). Ecosystem services – Examples of their valuation methods in Czech Republic and Slovakia. *Change and Adaptation in Socio-Ecological Systems, 1*(1). https://doi.org/10.2478/cass-2014-0008.

Schneider, J., Holušová, K., & Karber, V. (2016). *Proceedings from the role of forest functions within ecosystem services international scientific conference* (p. 2016). Brno: Mendel University in Brno.

Schröter, M., Albert, C. H., Marques, A., Tobon, W., Lavorel, S., Maes, J., Brown, C., Klotz, S., & Bonn, A. (2016). National Ecosystem Assessments in Europe: A review. *BioScience, 66*(10), 813–828. https://doi.org/10.1093/biosci/biw101.

Šefferová Stanová, V., & Galvánková, J. (Eds.). (2015). *Monitoring of plants and habitats of community interest in the Slovak Republic, results and assessment in the period of 2013–2015*. Banská Bystrica: State Nature Conservancy of the Slovak Republic.

Sharp, R., Tallis, H.T., Ricketts, T., Guerry, A.D., Wood, S.A., Chaplin-Kramer, R., Nelson, E., Ennaanay, D., Wolny, S., Olwero, N., Vigerstol, K., Pennington, D., Mendoza, G., Aukema, J., Foster, J., Forrest, J., Cameron, D., Arkema, K., Lonsdorf, E., Kennedy, C., Verutes, G., Kim, C.K., Guannel, G., Papenfus, M., Toft, J., Marsik, M., Bernhardt, J., Griffin, R., Glowinski, K., Chaumont, N., Perelman, A., Lacayo, M. Mandle, L., Hamel, P., Vogl, A.L., Rogers, L., Bierbower, W., Denu, D., & Douglass, J. (2018). *InVEST Version 3.4.4 user's guide*. The Natural Capital Project, Stanford University, University of Minnesota, The Nature Conservancy, and World Wildlife Fund. https://naturalcapitalproject.stanford.edu/. Accessed 10 Apr 2019.

Sherrouse, B. C., Clement, J. M., & Semmens, D. J. (2011). A GIS application for assessing, mapping, and quantifying the social values of ecosystem services. *Applied Geography, 31*, 748–760. https://doi.org/10.1016/j.apgeog.2010.08.002.

Spangenberg, J. H., & Settele, J. (2010). Precisely incorrect? Monetising the value of ecosystem services. *Ecological Complexity, 7*(3), 327–337. https://doi.org/10.1016/j.ecocom.2010.04.007.

Špulerová, J. (2006). Funkcie nelesnej drevinovej vegetácie v krajine. *Životné prostredie, 40*(1), 37–40.

Špulerová, J., Dobrovodská, M., Štefunková, D., Bača, A., & Lieskovský, J. (2014). *Biodiversity of traditional agricultural landscapes in Slovakia and their threats. Biocultural Landscapes.* Dordrecht: Springer.

Špulerová, J., Hrnčiarová, T., Piscová, V., Vlachovičová, M., Kalivoda, H., Kanka, R., Dobrovodská, M., Kenderessy, P., Miklósová, V., Drábová, M., & Belčáková, I. (2016). Sustainable tourism development in a selected area of the Low Tatras National Park – landscape planning versus urban planning. *Carpathian Journal of Earth and Environmental Sciences, 11*(2), 485–496.

Špulerová, J., Štefunková, D., Dobrovodská, M., et al. (2017). *Historické štruktúry poľnohospodárskej krajiny Slovenska.* Bratislava: VEDA.

Špulerová, J., Petrovič, F., Mederly, P., Mojses, M., & Izakovičová, Z. (2018). Contribution of traditional farming to ecosystem services provision: Case studies from Slovakia. *Land, 7*(2), 74. https://doi.org/10.3390/land7020074.

Štěrbová, M. (2017). Prístupy a metódy hodnotenia ekosystémových služieb lesa. *Životné prostredie, 51*(4), 213–220.

Stevens, M., Demolder, H., Jacobs, S., Michels, H., Schneiders, A., Simoens, I., Spanhove, T., Van Gossum, P., Van Reeth, W., & Peymen, J. (2015). *Flanders regional ecosystem assessment – State and trends synthesis report.* https://pureportal.inbo.be/portal/files/9004761/Stevens_etal_2015_FlandersRegionalEcosystemAssessment_State_Trends.pdf. Accessed 10 Apr 2019.

Supuka, J., et al. (1991). *Ekologické princípy tvorby a ochrany zelene.* Bratislava: Veda.

Supuka, J., Jančura, P., Pavlík, J., Kattoš, K., & Marenčák, M. (2000). *Ekológia urbanizovaného prostredia.* Zvolen: Technical University in Zvolen.

Švajda, J. (2009). Contribution for improvement of visitor monitoring in the Tatra National Park. *Eco. Mont-Journal on Protected Mountain Areas Research, 1*, 13–18. https://doi.org/10.1553/ecomont2s13.

TEEB – The Economics of the Ecosystem and Biodiversity. (2010). *Mainstreaming the economics of nature. A synthesis of the approach, conclusions and recommendations of TEEB.* Geneva: United Nations Environment Programme.

Turanovičová, M., & Rózová, Z. (2017). Metodické východiská hodnotenia kultúrnych ekosystémových služieb v urbanizovanom prostredí. *Životné prostredie, 51*(4), 232–239.

Turner, K. G., Vestergaard Odgaard, M., Bøcher, P. K., Dalgaard, T., & Svenning, J. C. H. (2014). Bundling ecosystem services in Denmark: Trade-offs and synergies in a cultural landscape. *Landscape and Urban Planning, 125*, 89–104. https://doi.org/10.1016/j.landurbplan.2014.02.007.

Tutka, J., Vilček, J., & Kovalčík, M. (2009). Oceňovanie verejnoprospešných funkcií lesných a poľnohospodárskych ekosystémov a služieb odvetví. In *Abstracts of Aktuálne otázky ekonomiky lesného hospodárstva Slovenskej republiky.* Technical University in Zvolen.

UK NEA. (2011). *The UK National Ecosystem Assessment: Synthesis of the key findings.* Cambridge: UNEP-WCMC.

Vačkář, D., Grammatikopoulou, I., Daněk, J., & Lorencová, E. (2018). Methodological aspects of ecosystem service valuation at the national level. *One Ecosystem, 3*, e25508. https://doi.org/10.3897/oneeco.3.e25508.

Value of Nature to Canadians Study Taskforce. (2017). *Completing and using ecosystem service assessment for decision-making: An interdisciplinary toolkit for managers and analysts.* Ottawa: Federal, Provincial, and Territorial Governments of Canada. https://biodivcanada.chm-cbd.net/documents/ecosystem-services-toolkit?lang=En& n=B443A05E-1.Accessed 10 Apr 2019.

Verweij, P., Janssen, S., Braat, L., Van Eupen, M., Pérez Soba, M., Winograd, M., De Winter, W., & Cormont, A. (2016). QUICKScan as a quick and participatory methodology for problem identification and scoping in policy processes. *Environmental Science and Policy, 66*, 47–61. https://doi.org/10.1016/j.envsci.2016.07.010.

Vilček, J. (2011). Potenciály a parametre kvality poľnohospodárskych pôd Slovenska. *Geografický časopis, 63*(2), 133–154.

Vilček, J. (2014). Mapovanie a hodnotenie environmentálnych funkcií poľnohospodárskych pôd Slovenska. *Geografický časopis, 66*(3), 287–304.

Vilček, J., & Koco, Š. (2018). Integrated index of agricultural soil quality in Slovakia. *Journal of Maps*, 14(2), 68–76 (2018). https://doi.org/10.1080/17445647.2018.1428233.

Villa, F., Ceroni, M., Bagstad, K., Johnson, G., & Krivov, S. (2009). *ARIES (Artificial Intelligence for Ecosystem Services): A new tool for ecosystem services assessment, planning, and valuation*. Burlington: Ecoinformatics Collaboratory, Gund Institute for Ecological Economics, University of Vermont. www.ariesonline.org/docs/Villa%20et%20al.%20 2009.Pdf. Accessed 10 Apr 2019.

Vološčuk, I. (2013). *Teoretické princípy ekologických procesov, funkcií a služieb ekosystémov*. Banská Bystrica: Belianum.

Wijnja, H., Van Uden, G., & Delbaere, B. (Eds.). (2016). *Ecosystem Services in operation: Case studies*. European Commission FP7, OpenNESS Project. https://issuu.com/ecnc.org/docs/openness_casestudies_brochure. Accessed 20 Mar 2019.

Zulian, G., et al. (2013). *ESTIMAP: Ecosystem services mapping at European scale*. Belgium: Institute for Environment and Sustainability, Joint Research Centre, European Commission.

Zulian, G., Stange, E., Woods, H., Carvalho, L., Dick, J., Andrews, C. H., Baró, F., Vizcaino, P., Barton, D. N., Nowel, M., Rush, G. M., Autunes, P., Fernandes, J., Ferraz, D., Ferreira Dos Santos, R., Aszalós, R., Aranay, I., Crúcz, B., Priess, J. A., Hoyer, C. H., Bürger-Patricio, G., Lapola, D., Mederly, P., Halabuk, A., Bezák, P., Kopperoinen, L., & Viinikka, A. (2018). Practical application of spatial ecosystem service models to aid decision support. *Ecosystem Services, 29*(C), 465–480. https://doi.org/10.1016/j.ecoser.2017.11.005.

Part II
Ecosystem Services of Slovakia

Chapter 2
Methodology of the National ES Assessment

Peter Mederly, Jana Špulerová, Zita Izakovičová, František Petrovič, and Ján Černecký

Abstract Chapter provides a brief overview of the methodology of pilot ES assessment in Slovakia. For the assessment process, in total 41 map inputs were used, in raster format with the pixel size 25 m. The most important data are the land cover map, the ecosystems map, protected areas, forestry data, digital elevation model, and soil data. The resulting landscape capacity maps present selected ES in the 0–100 relative scale, in a uniform standardized pixel format of 1 km resolution. Background data contain about 49,000 pixels with individual ES values and thus represent a basic dataset which is possible to use for further evaluation of the relationships and factors which affect ES provision.

The main aim of the publication is to provide a pilot assessment of all ES, which were selected for the territory of Slovakia based on current ecosystem and ES assessment process in Europe and the MAES process in Slovakia (5 provisioning, 10 regulating/supporting, and 3 cultural ES). As mentioned in Chap. 1, the issue of ES assessment is extremely complex and involves several aspects. Moreover, this process is still only in its beginnings within Slovakia and is not well elaborated.

P. Mederly (✉) · F. Petrovič
Faculty of Natural Sciences, Constantine the Philosopher University in Nitra, Nitra, Slovakia
e-mail: pmederly@ukf.sk; fpetrovic@ukf.sk

J. Špulerová · Z. Izakovičová
Institute of Landscape Ecology of the Slovak Academy of Sciences, Bratislava, Slovakia
e-mail: jana.spulerova@savba.sk; zita.izakovicova@savba.sk

J. Černecký
Faculty of Natural Sciences, Constantine the Philosopher University in Nitra, Nitra, Slovakia

State Nature Conservancy of the Slovak Republic, Banská Bystrica, Slovakia

Institute of Landscape Ecology of the Slovak Academy of Sciences, Bratislava, Slovakia
e-mail: jan.cernecky@sopsr.sk

P. Mederly, J. Černecký (eds.), *A Catalogue of Ecosystem Services in Slovakia*,
https://doi.org/10.1007/978-3-030-46508-7_2

Therefore, references from various available sources, in particular scientific reviews and articles, were used for the characterization of individual ES.

To improve the levels of knowledge and as an introduction to the next stage of the detailed and applied ES assessment in Slovakia, we consider it appropriate and useful to draw up a clear assessment of the main ES which we have tried to establish in this publication. We have developed the *coordinated procedure for expressing the landscape's relative capacity to provide each ES*, based mainly on biophysical (environmental) spatially expressed data. The result of the assessment is a relative scale of 0–100, where 0 means the minimum and 100 the maximum suitability of the area for the provision of the given ES within the whole territory of Slovakia (Mederly et al. 2019). The values can then be classified into a simple suitability scale, e.g., minimum to low–below average–average–above-average–high to very high capacity of the landscape to provide ES.

Background map documents and database information were prepared in a unified form to assess the landscape's capacity to provide ES. In particular, we relied on relevant data available in the spatial and information datasets of organizations involved in the compilation of the publication (Constantine the Philosopher University in Nitra, ILE SAS, and SNC SR), in some cases from the available sources of environmental institutions (SEA and SHMÚ – Slovak Hydrometeorological Institute). In total, 41 map layers were used directly for the ES evaluation (see Table 2.1), which were compiled into additional tailored assessment layers via reclassification or computational algorithms. Subsequently, the resulting capacity maps of individual ES were prepared. Key layers which were used for a larger number of ES included: a map of the current landscape structure and its interpretation, a map of ecosystems and the selected derived features, basic data on Slovakia's forests, data on protected areas, a digital elevation model, and soil properties data. The detail and applicability of these documents fit a scale of 1:10000 to 1:25000, which is highly above standard for national-level assessment. Supplementary background material was aimed at expressing important climate and hydrological data, where the accuracy is at a scale of approximately 1:50000, which is also sufficient for the national level. For cultural ES, less accurate documents from the Landscape atlas of the Slovak Republic were used – in order to ensure a more accurate assessment of the ES, a task remains to incorporate the information, especially at the level of individual municipalities, from state statistical surveys.

The standardization of the background layers consisted of converting all input maps into a uniform shape – a raster format with a pixel size of 25 m. All calculations were performed in this resolution, and also a landscape capacity map was generated to provide the given ES in different arithmetic values (generally from 0 to N points), based on the respective calculation algorithm different for each ES. For better standardization of results, their better display ability, and preparation for further statistical analysis, the capacity values were then recalculated for grid of 1 km – resulting value for pixel 1 × 1 km was calculated as an arithmetic average of values from 1600 original pixels 25 × 25 m. These values were then converted to a 0–100 scale according to a simple transformation algorithm, where 0 = lowest achieved value and 100 = highest achieved value for a given ES. *The resulting ES maps for*

Table 2.1 List of map layers used for assessment of ecosystem services in Slovakia

Content (theme) of the map layer	Source of data	Accuracy	Prov.	Reg.	Cult.
Digital elevation model – slope and other parameters	Database of UKF	1:25,000	2	2	2
Morphological-positional type of relief	Database of ILE SAS	1:25,000	*	1	2
Hydrogeological regionalization	Database of ILE SAS	1:50,000	1	*	*
Average annual temperature	SR Climate Atlas	1:50,000	*	1	*
Rainfall intensity (max 1-day totals)	SR Climate Atlas	1:50,000	*	1	*
Moisture balance indicator	SR Climate Atlas	1:50,000	*	1	*
Average. annual amount of solar radiation	SR Climate Atlas	1:50,000	*	1	*
Territorial climate classification	SR Climate Atlas	1:50,000	2	*	*
Hydrological basins (watersheds)	Slovak Water Mng. Map	1:50,000	*	1	*
Watercourses and water bodies	Slovak Water Mng. Map	1:50,000	1	*	*
Significant watercourses	Slovak Water Mng. Map	1:50,000	1	*	2
Water resources used	Slovak Water Mng. Map	1:50,000	1	*	*
Water resources protection zones	Slovak Water Mng. Map	1:50,000	1	*	*
Water reservoirs	Slovak Water Mng. Map	1:50,000	1	*	*
Basins of watercourses used for drinking purposes	Slovak Water Mng. Map	1:50,000	1	*	*
Natural medicinal resources protection zones	Slovak Water Mng. Map	1:50,000	1	*	1
Protected water management areas	Slovak Water Mng. Map	1:50,000	1	*	*
Average. groundwater depth	Database of ILE SAS	1:25,000	2	*	*
Soil subtype	SSCRI, ILE SAS	1:25,000	2	2	*
Soil texture	Database of ILE SAS	1:25,000	2	2	*
Soil depth	Database of ILE SAS	1:25,000	2	1	*
Current landscape structure/land use	ZB GIS, Corine Land Cover	1:25,000	3	3	3
Spatial diversity of landscape structure	Database of UKF	1:25,000	*	2	*
Classification and use of forest spatial units	NLC + SNC SR	1:10,000	2	*	2
Forest types	NLC + SNC SR	1:10,000	*	3	*
Forest age classes	NLC + SNC SR	1:10,000	1	3	1
Significant ecosystems (habitats)	SNC SR	1:25,000	*	2	*
The naturalness of ecosystems	Database of UKF	1:25,000	*	2	*
State of ecosystems	SNC SR	1:25,000	*	1	*
Categorization of protected areas	SNC SR	1:25,000	*	1	2

(continued)

Table 2.1 (continued)

Content (theme) of the map layer	Source of data	Accuracy	Prov.	Reg.	Cult.
Natural conservation significance of a territory	Database of UKF	1:25,000	*	1	2
Leaf area index (LAI)	Copernicus Global Land S.	1:50,000	*	2	*
Photosynthetically active radiation (FAPAR)	Copernicus Global Land S.	1:50,000	*	1	*
Normalized difference vegetation index (NDVI)	Copernicus Global Land S.	1:50,000	*	1	*
Potential for geothermal energy	SR Landscape Atlas	1:100,000>	*	*	1
Fishing and hunting areas	SR Landscape Atlas	1:100,000>	1	*	*
Areas of traditional (historical) land use	SR Landscape Atlas	1:100,000>	*	*	3
Significant natural sites	SR Landscape Atlas	1:100,000>	*	*	2
Historical parks and gardens	SR Landscape Atlas	1:100,000>	*	*	2
Cultural and historical attractions and monuments	SR Landscape Atlas	1:100,000>	*	*	2
Recreation and tourism objects	SR Landscape Atlas	1:100,000>	*	*	1

3 – most important layers for ES assessment, 2 – important layers for ES assessment, 1 – complementary layers for ES assessment; Prov. - Provisioning ES, Reg. - Regulatory & Supporting ES, Cult. - Cultural ES

the territory of Slovakia contain about 49,000 pixels with individual values for each ES – they represent a basic statistical set (or a point field) with which further work can be performed and assessment of the interactions and factors affecting the provision of the ES can be made.

Since the distribution of the majority of the resulting ES capacity values was significantly asymmetric and did not meet the preconditions for a statistically normal distribution, before the final transformation of the maps into the 0–100 scale, we proceeded to the modification – cutting the data file by so-called outliers at 2% of the minimum and maximum values. *The graphical presentation of maps* in the publication is unified – maps show the relative capacity of a landscape to provide a given ES in a 5-degree legend divided by the frequency of occurrence (i.e., every 20% percentile of occurrence is represented by 1 shade of a given color scale).

The resulting capacity of the landscape is not expressed in biophysical or monetary values, but in a relative scale (it represents % of maximum capacity, the value of suitability of the area, etc.). The big advantage is that these values can be further processed on the basis of known data from relevant research and studies. Minimum and maximum values can be replaced by specific biophysical units or monetary values based on advanced research or value/benefit transfer method from known ES valuation studies. It is this path that could be seen as a promising for the future assessment of the capacity of Slovakia's landscape to provide ES.

As part of the ES description, we also assess two factors of the spatial distribution of the ES in Slovakia – the relationship between the main types of landscape and ecosystems which provide the given ES; and the importance of the ES in terms of nature and landscape protection in Slovakia.

The main types of landscape and ecosystems providing the given ES are described based on the comparison of the spatial distribution of individual landscape types/main ecosystems with the achieved value of the landscape capacity. Most ES maintain a *logic* of the correlation between the degree of naturalness of a given landscape/ecosystem type and its ES provisioning capacity, but this is not always the case (e.g., for some provisioning services).

The importance of the given ES from the point of view of nature and landscape protection in Slovakia is assessed (both verbally and graphically) by comparison of the achieved capacity of the landscape with the degree of nature conservation significance of an area in the SR (I.–V.). *The nature conservation significance* is a special indicator used also for the processing of some ES maps, which expresses the synthesis of various existing categories of nature and landscape protection – from the national system of protected areas of the SR, through the European system NATURA 2000, biosphere reserves Man and Biosphere Programme (MAB), and the United Nations Educational, Scientific and Cultural Organization (UNESCO) natural heritage sites, to Ramsar Sites in Slovakia. Based on the overlap of individual categories of nature and landscape protection, we have compiled a more detailed 9-stage and simpler 5-stage classification of the nature conservation significance of Slovakia's territory (categories I–V), where I. represents an area without any protection and V. is an area with at least three overlapping categories of protection at the same time. These categories do not, therefore, represent degrees of protection 1–5 as per the Act on Nature and Landscape Protection of the SR – e.g., the area of protection level 5 without any other categories of protection represents the nature conservation significance degree III. The relationship between the categories of nature conservation significance and the landscape's capacity to provide ES was expressed in a graph and a simple correlation for each ES.

The *comprehensive* ES comparison according to their basic groups, the final assessment of the achieved results, their relation to nature conservation significant areas, and proposals for the further continuation of the ES assessment process in Slovakia are present in Chap. 6 of the publication.

Reference

Mederly, P., Černecký, J. (eds.) et al. (2019). Katalóg ekosystémových služieb Slovenska. Banská Bystrica: State Nature Conservancy of the Slovak Republic, Constantine the Philosopher University in Nitra, Institute of Landscape Ecology SAS.

Chapter 3
Provisioning Ecosystems Services

Peter Mederly, Martin Jančovič, Dominika Kaisová, Gréta Vrbičanová, Simona Gusejnov, Matej Močko, Ivan Laco, and Tomáš Kováč

Abstract Chapter provides the analysis and assessment of **five provisioning ES** – *P1 Biomass: Agricultural crops; P2 Biomass: Timber and fibre; P3 Drinking water; P4 Freshwater; P5 Fish and Game/Wildfood.* All ES are described in the unified structure: definition and brief characteristics; methods used for identification and assessment; main types of landscape and ecosystems providing given ES; the importance of ES in terms of nature and landscape protection; and ES assessment for the territory of Slovakia. Spatial assessment is provided as a map of the landscape capacity for given ES provision. For all ES, short conclusions and overview of input data for further assessment of the ES capacity, demand and flow are also given.

3.1 Biomass: Agricultural Crops (P1)

3.1.1 Definition and Brief Characteristics of ES

The production of biomass for food is the basic provisioning ES – it is provided by different types of landscape, among which the most important are *agroecosystems*. In addition to the production of food and related raw materials, these ecosystems also provide another provisioning ES, such as fodder and energy biomass; they also

P. Mederly (✉) · M. Jančovič · D. Kaisová · G. Vrbičanová · S. Gusejnov · M. Močko
I. Laco · T. Kováč
Faculty of Natural Sciences, Constantine the Philosopher University in Nitra, Nitra, Slovakia
e-mail: pmederly@ukf.sk; martin.jancovic@ukf.sk; dominika.kaisova@ukf.sk; greta.vrbicanova@ukf.sk; simona.stasova@ukf.sk; matej.mocko@ukf.sk; ivan.laco@ukf.sk; tomas.kovac@ukf.sk

P. Mederly, J. Černecký (eds.), *A Catalogue of Ecosystem Services in Slovakia*,
https://doi.org/10.1007/978-3-030-46508-7_3

play an important role in other regulating and supporting ES. Agricultural landscape provides people space in which food is produced, people can live and the landscape serves as a place for agro-tourism and recreation, providing cultural heritage and aesthetic values, including various cultural ES (Schröter et al. 2019). To a lesser extent, this ES is also provided by seminatural and natural ecosystems.

According to Preston and Raudsepp-Hearne (2017), crops are plant products which people need for bio-nutrition or commercial use (except for field crops such as fruits, seeds, vegetables and herbs). In particular, ecosystems provide soil, nutrients and microbiological and climatic conditions which allow people to cultivate food – through the natural gross primary production and the conversion of solar energy into biomass, partly by energy transfer in food chains and water and nutrient cycles.

Agriculture crop production is one of the key ES for food provision. It depends on a number of factors – from natural (soil quality, climate, water availability, pollination) through socio-economic (labour availability, food demand) to purely economic (macroeconomic, market relations). The productivity and efficiency of plant production are dependent on cropland extent, yields and crops produced. The actual yield produced also depends on other factors, such as the genetic potential of the crop; the amount of solar radiation, water and nutrients absorbed by the crop; the presence of weeds and pests; etc. (available online: www.data.oecd.org). However, crop and biomass production is mainly dependent on ecosystems and their ES. In turn, agricultural ecosystems contribute to other ecosystem services not related to biomass and crop production (Power 2010).

ES Biomass – crop production can be understood as *the ability of ecosystems to provide food provision services*. This ES is mainly dependent on the agricultural ecosystems and their surroundings, which determine the overall ES production capacity in qualitative and quantitative terms. To a large extent, it is linked to other provisioning ES providing additional raw material resources related to biomass and crop production.

3.1.2 Methods Used to Assess and Identify ES

For ES Biomass – agricultural crops, there are different assessment methods which are considered suitable and which are applied. Consideration is given to, for instance, inputs for production (characteristics of agroecosystems providing agricultural production) or direct production (yields of individual crops, sales for production, etc.). In particular, *economic and biophysical assessment methods are used*.

In general, the most used indicators of the assessment of this ES include crop areas and crop and fruit yield (Maes et al. 2014; Czúcz et al. 2018); other indicators used are soil production properties (biomass production, nutrients), climatic parameters, but also, for instance, water quality for irrigation (Pérez-Soba et al. 2015). In some cases, national ES assessments are based on the country's regional statistics on crop production (Denmark, Romania) or from data on the area of agricultural land, arable land and crop areas of individual crops (Luxembourg, Romania). Several assessments are based on the amount of biomass produced, crop yields, soil fertility, taking into account the climatic conditions and calculating various environmental performance and stability indices (Germany, Romania). The assessment of this ES is based, for example, on a simple model of average production for a particular type of crop, taking into account its economic value. The production of a particular crop in a given area (e.g. in kg/ha/year) is determined, which, when converted to the average market value of the crop grown per area, expresses its economic value (UK, Spain). In the national ES assessment of Spain, this outcome was compared with the outputs of the High Nature Farming project, where the relations between the economic and natural value of the agricultural landscape were investigated (Santos-Martín et al. 2016).

One of the tools of the biophysical assessment of ES Biomass – agricultural crops is the InVEST, which is a modern and relatively widespread ES assessment tool. Based on various input parameters, InVEST assesses the capacity and flow of different ES and is applicable in different areas and spatial scales, based on standard GIS methods. The tool contains a number of models, including the *crop production model*. It has been designed to help answer a number of fundamental issues related to agricultural productivity, e.g. what is the impact of different agricultural approaches and measures, crop rotation and intensification of production on the provision of this ES, and how the growing demand for food can be met with minimizing the impact on other ES. Input data for the percentile model operation is formed by a data set that includes information on 175 crops (included in the model installation). Furthermore, there is a need for data on the current landscape structure of the surveyed area, with data on cultivated crops. In addition, for a regression model, fertilizer rate data for each crop should be entered – but this model only works with 12 basic crops. The output of the model includes the data on production volume, nutritional values and production rate of all modelled crops (available online: www.naturalcapitalproject.org).

An interesting assessment may also be provided by the application of a *matrix model* which determines the basic parameters of this ES for agricultural land. Consequently, it is possible to combine these data with the state of protection of individual natural ecosystems in the immediate vicinity, which provide microclimatic and soil protection functions for the given area and also provide space for natural pollinators. Such an assessment will ensure a better understanding of the interactions of intensification of agriculture in contrast to the interconnection of agricultural land to natural ecosystems and their functions, as well as mutual interactions. In basic models, these facts are often omitted.

The production potential of agricultural land in Slovakia is expressed numerically by means of a scoring system for the so-called Soil-Ecological Units (SEU) system (Džatko 2002). The maximum score is 100, the minimum is 2. Vilček (2011) mentions soil assessments in terms of energy potential or energy accumulation. From these statistics and methods of assessment, it is possible to build a further assessment of the soil and agricultural land ES in Slovakia.

3.1.3 The Main Types of Landscape and Ecosystems Which Provide ES

The basic landscape component in terms of the production of agricultural biomass is *soil* as an essential part of agricultural ecosystems (so-called agroecosystems), with another provisioning ES also depending on its quality. Of course, for the proper functioning of the agricultural landscape ecosystems and for crop yields, the favourable condition of other landscape components, their properties (e.g. climatic conditions, state and regime of surface and groundwaters, the occurrence of disturbances) and human activity are also important (Bezáková 2015).

Agroecosystems represent altered ecosystems which have been modified for the purpose of producing food and biomass for use – fibres (Hodgson 2012). A typical agroecosystem contains 1–4 major crops and 6–10 major pests. Typically, an agroecosystem includes less diversity of animal and plant species than a natural ecosystem (forest, meadow) (Karuppuchamy and Venugopal 2016). The intensively used agroecosystem is subject to rapid changes due to anthropogenic influences such as ploughing, fertilization and application of pesticides (Fig. 3.1). While retaining many characteristics from natural ecosystems, from the toxicological point of view, it is characterized by an increased presence of agrochemicals, including pesticides, fertilizers and plant growth regulators (Hodgson 2012).

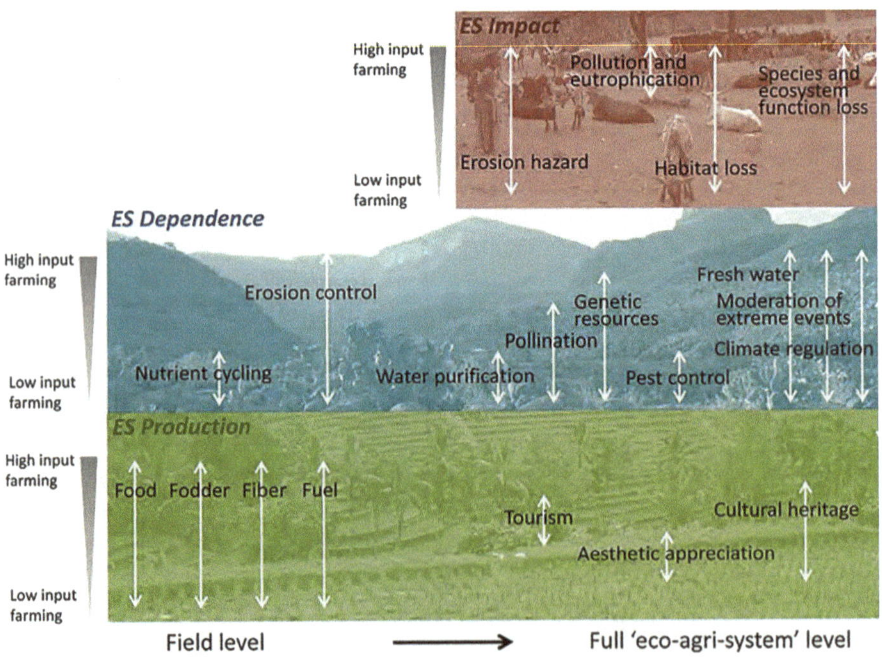

Fig. 3.1 The relationship of agricultural management towards ecosystem services. (Source: Burkhard and Maes 2017)

According to Burkhard and Maes (2017), the agroecosystems producing agricultural crops represent the most important ecosystems for man in modern epoch – Anthropocene. Up to 44% of Europe's land surface is agricultural land. Agroecosystems have a significant impact on the environment, including ecological processes and functions.

ES Biomass – agricultural crops uses most of the total area of agroecosystems. In terms of the structure of the agricultural lands, this ES is based mainly on arable land (growing of cereals, legumes, oilseeds, root crops), but it also includes fruit growing, hop growing, vegetable growing and viticulture. Therefore, the *agroecosystems of permanent crops* – especially gardens, orchards and vineyards – are very important elements of the landscape structure associated with this ES and special agroecosystems. These increase the diversity of the agricultural landscape and often also represent the remnants of historical land use elements and providers of other ecosystem functions and services such as crop production. Indeed, modern intensive farming is mainly focused on the production function of agricultural land, which largely suppresses other ES, in particular regulating services and supporting functions, partly cultural services.

As regards the other types of ecosystems, they are not directly linked to the production of crops. Nevertheless, some of them may have a positive impact on it – especially aquatic ecosystems and wetlands (maintaining water availability for the agroecosystem), scattered vegetation in the country (e.g. pollination support, regulation of threatening processes) and partly forest ecosystems (e.g. influencing local climate and hydrological regime). Even some anthropogenic ecosystems support, to some extent, the plant production, or they make it more efficient (road network, agricultural farms) but do not provide the given ES separately.

3.1.4 Importance of ES in Terms of Nature and Landscape Protection in Slovakia

It has been proven that many agricultural practices and the expansion of agriculture pose a *major threat to the proper functioning of ecosystems*. On the other hand, well-managed agriculture can be an important means of securing and protecting ecosystems and the ES (Burkhard and Maes 2017). Thus, as far as the relationship between agricultural production and nature and landscape protection is concerned, in most cases it is perceived negatively (in accordance with Fig. 3.2) – the expansion and intensification of agriculture is the cause of the loss of environmental diversity and the quality of other types of ecosystems in different world regions. In Slovakia, this was particularly the case in the phase of collectivization and intensive socialist agriculture, but even today's European agriculture based on the system of agro-subsidies is not favourable in Slovakia in terms of the real ecological condition of the landscape (preference of certain crops, preservation of large blocks of agricultural land, excessive application of chemicals, *formal* care for the landscape).

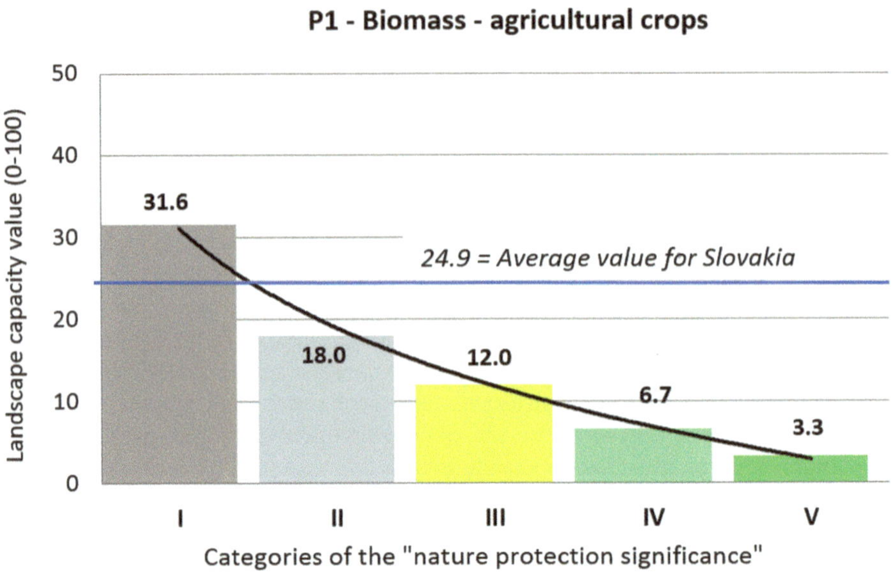

Fig. 3.2 Relationship of ecosystem service P1 and significance of Slovakia territory in terms of nature and landscape protection

Fig. 3.3 Agricultural landscape with important ecosystem elements for animals and raptors (surroundings of Suchá nad Parnou). (Source: D. Štefunková)

Nature and landscape protection itself have targets which are completely different from food production. However, as mentioned earlier, the *support of biodiversity and the ecological quality of the landscape* can indirectly have a positive impact on the production function of agroecosystems (Fig. 3.3) – therefore this relationship may not necessarily have been completely antagonistic. Sufficient food needs to be produced in a sustainable way, taking into account challenges such as climate change and the growing population with changing dietary habits. Maintaining biodiversity in agroecosystems (the diversity and variability of animal species, plants and microorganisms) is important for food production and for preserving the ecological basis necessary for maintaining rural life (available online: www.fao.org). The positive fact is that in the period from 2000 to 2015, the area of land in ecological agriculture production in Slovakia recorded a threefold increase – in 2015 a total of 186.5 thous. ha of land was farmed in this way (available online: www.enviroportal.sk), which represents 7.8% of the total agricultural land area.

The important thing is that the so-called agri-environmental schemes currently preferred in the EU Common Agricultural Policy are also based on the *support of non-production functions of agriculture*, which are in line with the nature and landscape protection objectives in Slovakia (more in detail on this in the chapter of *ES R10 (soil formation and composition maintenance)*. However, their implementation and realization is often controversial, and rules need to be better defined to address the ecological requirements of each species and habitats.

3.1.5 ES Assessment for the Territory of Slovakia

Food production is one of the most important ES in the SR as it is essential for the survival and appropriate nutrition level of the population. Although, it is a fact that at present the agricultural land seems much less important than in the past when local and regional food production was driven by the survival of the majority of the population. This is proven by the fact that the area of agricultural land in Slovakia has been decreasing continuously since 1990 and most significantly in the case of arable land (Pazúr and Bolliger 2017). In the period from 2000 to 2015, 2.1% of the area of agricultural land (510.5 km^2) was lost in Slovakia – the average annual decrease is more than 3400 ha, which means two cadastral areas of average municipalities. At present, agricultural land represents 48.6% of the SR area (23,819 km^2) – of which 59.1% is arable land (14,087 km^2), 35.9% permanent grasslands, 3.2% gardens, 1.1% vineyards and 0.7% orchards.

Nevertheless, the production of some agricultural crops has increased over the last decade, probably due to the intensification of land use and increased inputs into processing. Cereal production has increased, but from the point of view of the ES assessed, it is not positive that the production of oilseeds and technical crops has increased much, while the production of some food crops (especially potatoes and legumes) has decreased.

Typical areas with predominant agroecosystems for food production in Slovakia include the Slovenska Podunajská, Východoslovenská and Záhorská nížina and southern Slovakian basins, although food crops are also grown in cooler areas, especially in basins and sub-mountain areas (Fig. 3.4). Growing of fruits and vegetables is dominant in warm areas of Slovakia; the cultivation of grapes is typical in warm hillsides of low mountains and hills in the south of Slovakia (world-famous is the vineyard region of Tokaj, extending into Zemplín Hills).

There is a lot of available data to assess *the food production in the SR* – from national statistics of individual crops production through statistics at the level of regions to some data at the level of districts and farms (available online: www.statistics.sk). However, other data than just the production, or economic data are also important to determine the landscape's ability to fulfil this ES. For instance, Vilček and Koco (2018) published an integrated soil quality index for agricultural land in

Fig. 3.4 Cereal field in dispersed settlements near Detva. (Source: J. Černecký)

Slovakia, in which, in addition to production parameters, they took into account environmental parameters and threats.

When assessing the landscape's capacity of food production, it is appropriate to base it mainly on data on the natural components – as a rule, economic factors such as real agricultural production infrastructure are not taken into account in the capacity of a landscape. The key factor is the *fertility (production ability) of soils*, which is well and long-term documented in Slovakia on the basis of the SEU system (Soil Science and Conservation Research Institute SSCRI Bratislava – available online: www.vupop.sk). The main factors of soil fertility include soil depth, soil texture, soil skeleton, the dominant paedogenetic process represented by soil subtypes. To some extent, the SEU system also includes the characteristics of other important natural factors, which can also be evaluated on the basis of more precise background information – in particular, this includes the *climatic conditions* (temperature and moisture balance), *relief characteristics* (slope, predisposition to erosion processes) and *geological substrate* (mineral richness, suitability for plant growth, hydrogeological conditions). It is possible to assess the overall production potential of the agricultural landscape or the potential for growing the most important, or selected crops.

The pilot assessment of the landscape's capacity to provide ES P1 was based on two main input indicators and three supplementary indicators (Table 3.1). All the indicators were spatially expressed, and a map of the landscape's capacity was compiled by combination based on the determined calculation algorithms. The resulting values were converted from the original resolution (pixel size 25 m) to 1 × 1 km spatial units and converted to a unified 0–100 scale, where 0 expresses the landscape's lowest capacity to provide this ES (infertile areas, built-up areas, water

Table 3.1 Input data for capacity, demand and flow assessment of ES Biomass – agricultural crops

Input data/ ES	P1 Biomass – Agricultural crops
Capacity	Current landscape structure – categories of agricultural land use and their suitability for growing crops
	Production potential of the SEU (agricultural land) and forest soils
	Slope inclination – reclassification, suitability coefficient 0.25–1.25
	Climatic conditions – temperature and moisture balance
	Hydrological conditions – availability (surface depth) of groundwater
Demand	Population – map of spatial distribution, population density
	Statistical data – consumption of selected food products (municipalities, districts)
Flow	Current landscape structure – categories of agricultural land use
	Statistical data – production of selected crops per area unit (kg/ha)
	Data on production of selected crops (municipalities, districts), conversion to the spatial unit

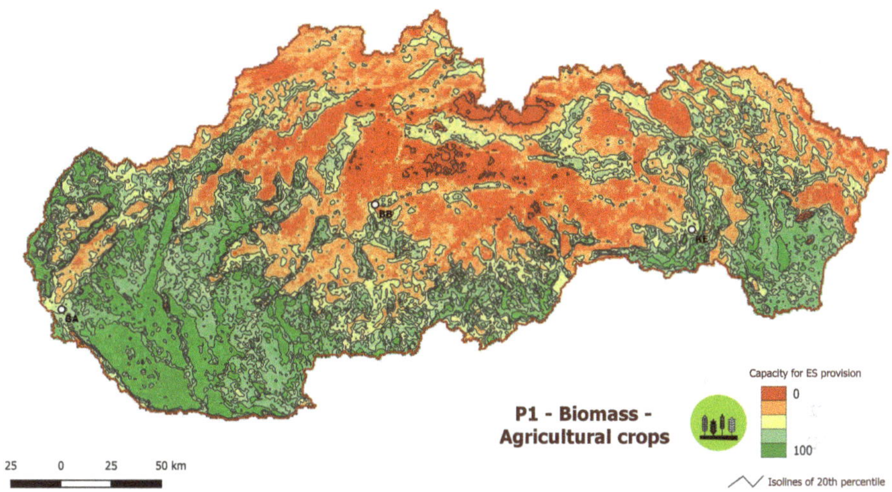

Fig. 3.5 Landscape capacity for provision of ES Biomass – agricultural crops (P1)

areas) and 100 theoretically the highest possible level of capacity (the most fertile soils on a flatland with the most favourable climatic and hydrological conditions). The results are shown in Fig. 3.5, which illustrates ES P1 compliance with the overall agricultural landscape productivity value.

For a comprehensive assessment in the future, the issue of the overall balance of the ES needs to be solved. This should be based on refining the value of capacity, determining the values of demand and real ES flows within the territory of Slovakia. The level of demand for this ES is clearly determined by the number of residents living in a particular territory, combined with the registered or anticipated consumption of different types of food of plant origin. The real use (flow) of this ES is, in turn, given by the number of crops and materials for food of plant origin produced in a particular territory that can be obtained from statistical surveys. However, the problem may be that, while assessing capacity data is appropriate using the natural regions, the ES demand and flow indicators mostly refer to administrative units (districts, municipalities). Therefore, it will be necessary to combine different spatial units in the future.

3.2 Biomass: Timber and Fibre (P2)

3.2.1 Definition and Brief Characteristics of ES

One of the main provisioning ES is the production of biomass, which can be further subdivided into specific subgroups, including timber and fibre production. Among the many ES provided by forests and which are essential for human benefits, the greatest financial importance is attributed to timber production. Timber serves as a raw material for the processing industry, construction or various technical purposes – e.g. as a cellulose source (MEA 2005). In simple terms, timber production can be defined as a *wide range of products derived from timber harvesting mainly from forest ecosystems or plantations* (Maes et al. 2013) as well as from trees outside the forest. In this process, forest ecosystems can be managed in different ways (intensively or extensively), and plantations may include forest or agricultural plantations. In literature, a distinction can be made between natural, seminatural, artificially planted forests and plantations (e.g. FAO).

In addition to timber, the processing industry also uses many other crops to obtain fibres. Thus, this provisioning ES can be understood to mean the cultivation and harvesting of fibres from agricultural crops (worth mentioning is cotton, flax, hemp or jute) or animal sources (this includes in particular wool, mohair or silk, but also various hides and skins; MEA 2005) for the production of, e.g. clothing, fabrics or paper (Kandziora et al. 2013). Fibres can also be obtained from timber – this includes, e.g. artificial silk (*rayon*) or lyocell.

The production of timber and fibres has historically been a particularly significant ES and still plays a crucial economic role nowadays, as reflected in the management of the ecosystems providing it. According to The Economics of Ecosystems and Biodiversity study (TEEB 2013), the production of timber and non-timber products (including fibres) is the primary economic function of up to 34% of the world's forests. With regard to fibres derived from agricultural production, with the exception of cotton and silk, their production has a predominantly downward trend (MEA 2005).

In summary, this provisioning ecosystem service can be defined as the production and collection of selected plant and animal resources predominantly for technical purposes or as the manufacturing input of the processing industry (timber, textile, paper, etc.).

3.2.2 Methods Used to Assess and Identify ES

In the context of Slovakia, the provisioning ES of production of biomass – timber and fibres – is partly considered when assessing the production functions of the forest, namely, the function of timber production. The value of timber production function of forests as a natural factor of production is usually based on market prices of timber products reduced by logging and transport costs (market method), but other methods of economic evaluation (available online: www.forestportal.sk) can also be used.

At the European level, several indicators have been identified in the context of the ES mapping and assessment process for the provision of biomass by forest ecosystems. These indicators include timber stock and increment, timber production (industrial, fibre timber, etc.) and timber consumption (e.g. logs), usually measured in m^3 (European Commission 2014). In case of biomass in the form of fibres, which is primarily provided by agricultural ecosystems, the indicators include crop yields which serve as a source of fibres crops (measured as, e.g. t/ha; t dry matter/ha; MJ/ha) or the area on which these crops are grown (ha) (European Commission 2014). It is a combination of indicators of landscape capacity, demand and the real use of this ES.

In the ES national assessments in European countries, the values of timber stock, forest growth and sustainable production levels (Germany, Romania, Russia) were mainly considered for this ES. Extraction and timber consumption were also assessed, but they are more about real flow and use of the ES.

For the incorporation of this ES into the natural capital accounting system (UN SEEA 2014), it is proposed to define timber resources by using the volume of timber biomass (including dead trees) and to include all trees regardless of their trunk circumference, as well as parts above the trunk, together with deadwood. Smaller branches and twigs, fallen leaves, flowers, seeds and roots are not included.

The above-mentioned indicators serve for *biophysical assessment* of this ES, whereby data can be obtained directly and expressed in biophysical units. However, in some cases, these data are not available or sufficient, so it is advisable to use indirect methods of measurement (remote sensing, statistical data) or modelling (e.g. crop simulation models such as Forest Vegetation Simulator) (Vihervaara et al. 2018; Binder et al. 2017).

Timber and timber products as well as fibre agricultural resources are generally traded commodities on specific markets, with significant economic value, which means that their *economic assessment* is done through a market mechanism (Binder et al. 2017). The value of the forest production function as a natural production factor can be determined on the basis of the market prices of the timber products reduced by the costs of logging and transport. However, other evaluation methods, such as value transfer (transfer of values from one study to another), substitution costs or contingent valuation, can be used (Forest Europe, available online: www.foresteurope.org/overview-valuation-approaches-methods/).

As far as *sociocultural assessment* is concerned, the production of biomass (timber and fibres) as such is often not assessed by these methods. This does not mean, however, that they cannot also be part of the assessment – these include various deliberative or participatory approaches, using, e.g. semi-structured questionnaires, group discussions, Q-methodology and others. These can be particularly beneficial, for example, when assessing people's preferences for the so-called trade-offs in the use of the ES – that is, the provision of one service (often provisioning) at the expense of the other (regulating, cultural) (DEFRA 2007; De Meo et al. 2018).

3.2.3 The Main Types of Landscape and Ecosystems Which Provide ES

In general, the production of biomass in the form of timber and natural fibres is mainly bound to *forest ecosystems* (forest stands, plantations and non-forest woody stands). Managed forest stands and plantations are used economically in particular. Forest stands differ in terms of production – their productivity depends mainly on the habitat conditions under which a particular type of forest community develops and of course the way of farming and care for the forest.

Forests are extremely important ecosystems in virtually all types of landscapes as they fulfil many functions and are involved in the provision of the full range of ES. Although their production function is undoubtedly very important to humans, its one-sided preference and use means depriving the landscape of other functions and services fulfilled by natural and seminatural forests (Fig. 3.6). The long-term economic use of forest stands has resulted in a gradual decline in natural forests (which are not the most suitable for economic use) and their replacement for more productive managed forests and plantations. In Slovakia, this mainly includes the substitution of beech and fir-beech forests for spruce monocultures and stands with a significant proportion of spruce or planting poplar plantations instead of the original floodplain forests.

According to summary information on the state of forests, the area of forest land in Slovakia reached 20,190 km^2 in 2017 (41.3% of the total area), of which 19,460 km^2 belongs to forest stands. Since 2000, the area of forest land increased by 180 km^2. At the same time, there is a long-term trend of increasing the area of forests and forest stands in Slovakia, but due to the ever-growing timber harvesting, the representation of higher age classes of forests is decreasing with the increase in young forests and harvested areas (MPSR NLC 2018).

Biomass is also provided by non-forest woodland ecosystems – in addition to hedges, e.g. also orchards, woody plants growing along roads and line structures or in urban parks (Binder et al. 2017). These types of stands can be described as an

Fig. 3.6 Lichens as important bioindicators of forest ecosystem quality and P2 ecosystem service. (Source: J. Černecký)

additional source of biomass in less forested areas with other main functions. A negative phenomenon at present is that many trees in the landscape are disappearing just because of the use of timber, and the landscape is thus deprived of their significant regulation and sometimes cultural-historical functions. However, statistical data on biomass recovery from these sources are not available. The agricultural policy of subsidies forces farmers to remove almost all the biomass from grassland habitats, thus losing important ecotone elements from the landscape, losing the transitional border between forest and non-forest habitats and decreasing the number of solitary trees. Such an approach is negative from the point of view of protection of many important types of agricultural landscapes and also from the viewpoint of ecological stability of the territory.

Natural fibres are also provided to a lesser extent by the agricultural ecosystems (cultivation of technical crops), but in Slovakia, it is rather marginal. Growing representation of flax plants (mainly flax and hemp) is negligible – as in 2018, only 439 ha were cropped (0.03% of arable land area).

For other ecosystems, provision for ES production of biomass is limited, e.g. some grassland ecosystems and wetlands (especially reed stands).

3.2.4 *Importance of ES in Terms of Nature and Landscape Protection in Slovakia*

The production of biomass (timber and natural fibres) is an important factor in terms of nature and landscape protection in Slovakia, as the provision of this ES competes with other services (e.g. water-flow and erosion regulation, biodiversity support, recreation and tourism) and raises the so-called trade-offs (providing one service at the expense of another). This ES is mainly provided by forest stands and woody plants outside the forest – they are currently expected to provide a wide range of ES, but in fact, significant conflicts of interest have been identified (Bradford and D'Amato 2012). One-sided management of forests to maximize timber production or carbon sequestration often results in a *negative impact on biodiversity* (Duncker et al. 2012) and thus on nature and landscape protection.

In Slovakia, forest management is governed primarily by Act no. 326/2005 Coll. on Forests as amended. Forests in Slovakia are divided into three basic categories: commercial, protection and special purpose. Timber production (while providing other important environmental and social services) is mainly intended for commercial forests, support for which is provided by specific forestry measures under the so-called function-integrated forestry. The focus of commercial forests on timber production is limited by the suitability of natural conditions, but they still represent the most represented forest category (72.1%) (MPSR NLC 2018). Protection forests and special-purpose forests are designated for the protection of nature and provision of other ES (water management, erosion protection, etc.), with priority to support non-provisioning ES (regulatory, supporting and cultural).

The analysis of the relationship between the nature conservation significance of the landscape in Slovakia and its capacity to produce timber (Fig. 3.7) shows that the landscape in the II.-IV. significance category has a high potential. It is in these territories that commercial forests prevail over other land use categories. Agricultural and urbanized landscape dominates the I. degree, with rare ecosystems being predominant in the V. degree. These rare ecosystems are either not part of the forest or may include significant representation of protection forests.

Nevertheless, Slovakia's problem is that commercial forests are often also located in protected areas, causing conflicts in relation to the main function which such territory should perform (provisioning function vs. nature protection). These conflicts can be minimized through appropriate management which takes into account the provisioning of multiple ES. For this purpose, modelling of the impacts of different management methods on ES provision may also be used (Carpentier et al. 2017). However, it is clear that consistent conservation of nature and landscape is a kind of *brake* for the intensive use of ES Biomass production, which is already inherent in its core principles.

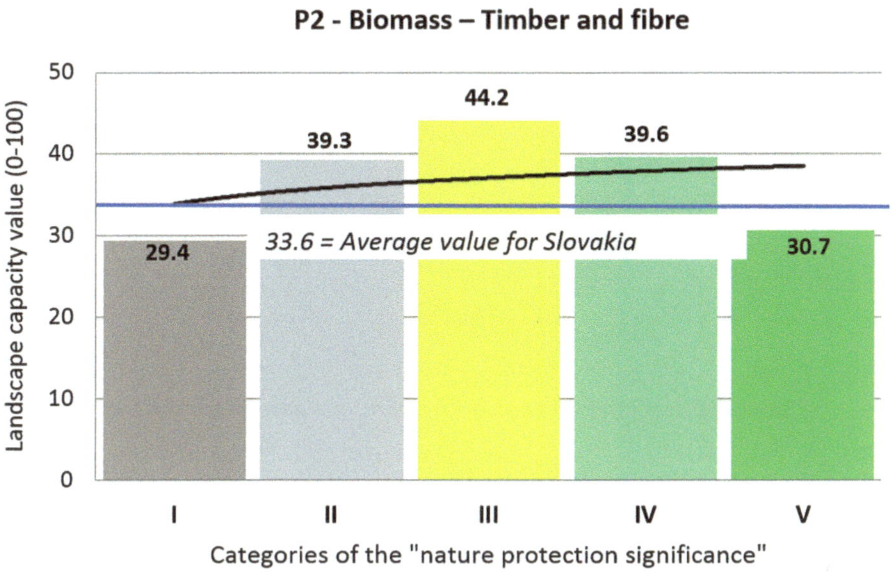

Fig. 3.7 The relationship between ecosystem service P2 and the significance of the territory of Slovakia in terms of nature and landscape protection

Significant conflicts between the use of the provisioning ES and the non-production functions and services of the landscape, including nature protection, occur in the case of biomass harvesting (Fig. 3.8 – especially for energy purposes) as well as in valuable types of non-forest ecosystems such as scattered vegetation in farmland, riparian vegetation and partly wetlands. This has been present mainly in recent years, paradoxically, because of the large subsidies for renewable energy sources and firewood.

3.2.5 ES Assessment for the Territory of Slovakia

As mentioned above, the amount of data for the assessment of ES Biomass production for conditions in Slovakia is sufficient, but only for forest stands. The volume of biomass (timber stock) and its use (timber harvesting) are regularly evaluated in the framework of the *Forest Care Program* for each decade and are statistically monitored. An important indicator is the *average increment*, which expresses the potential for increasing the volume of timber for a given period. Data on reserves and use of non-forest biomass is scarce, with the exception of statistics on the cultivation and yield of agricultural technical crops.

Wood harvesting in Slovakia has a long-term growing trend – while the total timber harvesting was 6218 thous. m^3 in 2000, in 2015 this number reached 9250 thous. m^3 and 9390 thous. m^3 in 2017 (50% increase). A negative phenomenon is that a very high proportion of this harvesting consists of the so-called random (or

Fig. 3.8 Example of intensive harvesting in the Nízke Tatry in spruce monoculture, which is susceptible to windthrow disaster or bark beetle outbreak. (Source: J. Hreško)

unplanned) harvesting – at an average level of 50–60%. Concerning the use of biomass, e.g. for heat production, in 2017 there was a 1425 thous. t of dendromass produced in the SR, which was three times more than in 2000 (476 thous. t). Almost the entire increases in the volume of production are represented by wood pellets (available online: www.enviroportal.sk).

The assessment of biomass production in the form of timber can be based on the procedure proposed for protected areas of Slovakia in the publication by Považan et al. (2014b). The authors propose to assess the production of timber using market prices, using a relatively simple formula.

$$Vt_a = St_a \times H_a \times Pt_a$$

where Vt_a is the value of the timber (EUR) produced in course of year a, St_a is the size of the area (ha) on which the timber has been harvested, and H_a is the average of timber harvesting (m^3/ha) per year, and Pt_a is the price of timber (EUR/m^3) in course of year a. Data from forestry enterprises, or national and regional statistics, can be used as a source of information. In case of insufficient sources of information on the amount of timber and sales prices for a detailed calculation at the level of forest reports or municipalities, it is possible to use the average harvesting and the average (mean) value of the timber price at regional or national level as a basis for such calculation. However, these indicators are representing the real use of this ES and do not present the relationship of timber harvesting with the natural potential and natural capital of the area.

If we want to assess the provision of ES production of biomass comprehensively, it is necessary to assess the indicators of the landscape's capacity to provide this ES, the demand for this service and its real use (flow). The capacity (potential) of the landscape for biomass production can be expressed in the case of forest stands by an indicator of *timber stock and increments*; in the case of agricultural crops, it can be expressed in the same way as the production of food crops (e.g. based on the production of agricultural land). Other biophysical indicators for forest stands may be represented by, for example, the *tree species quality*, expressing the production ability of woody plants or whole stands, or also the *production capacity of forest soils* (it is a similar indicator as in the case of agricultural soils).

A value transfer method can also be used for assessment, whereby the average price of timber (e.g. published for the EU) is based on the average timber production per hectare of area.

The pilot assessment of the landscape's capacity to provide ES P2 was carried out for the production of forest biomass or biomass from woody plants. Two basic and three additional input indicators were used (Table 3.2). The procedure was similar to that of ES P1. In the final map, the lowest capacity of the landscape for the provision of this ES is represented by built-up areas, water areas and infertile soils; the highest possible level of capacity (theoretical value 100) is achieved by the most fertile forest soils in favourable relief and climatic-hydrological conditions. The results are shown in Fig. 3.9. It is not surprising that mountain areas which are now almost continuously forested – especially the lower and medium-high mountain ranges – provide the largest capacity for the provision of this ES. The lowland and basin areas include more significant areas with higher representation of forests and non-forest timber vegetation.

Table 3.2 Input data for capacity, demand and flow assessment of ES Biomass – timber and fibre

Input data/ ES	P2 Biomass – timber and fibres
Capacity	Current landscape structure – selected categories of land use and their suitability for the production of timber and natural fibres
	Production capacity of forest stands (according to stand types in JPRL)
	Soil production potential – reclassification based on subtype, depth and texture
	Climatic conditions (temperature and moisture balance)
	Hydrological conditions (depth of groundwater level)
Demand	Statistical data – biomass consumption – timber and other fibres (municipalities, districts)
	Population – map of the distribution of population or population density (municipalities, districts)
	Value of commercial production (districts, regions)
Flow	Current landscape structure – CLS categories actually used to produce biomass (commercial forests, other trees, part of arable land, etc.)
	Statistical data – production of forest biomass, e.g. per unit area (kg/ha), by forest and agricultural enterprises
	State of forest ecosystems and measures in stands
	Size of crop areas, production of technical crops (municipalities, districts)

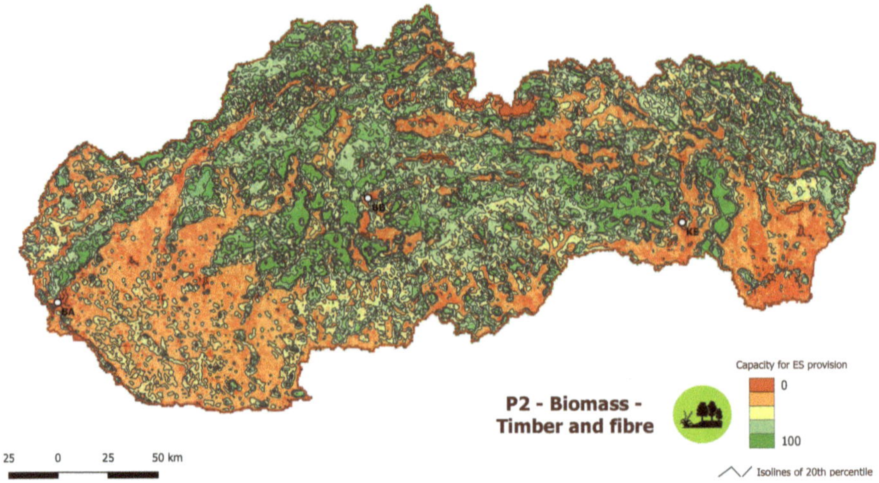

Fig. 3.9 Landscape capacity for provision of ES Biomass – timber and fibre (P2)

In addition to clarifying the value of the landscape's capacity for the provision of this ES, it is also necessary to address demand and actual use issues in the future. Demand for the use of ES Biomass production can be expressed by indicators of *real consumption of timber and other fibres* in different territories (administrative units), but such data is not available. Therefore, this data can be replaced by other related data – e.g. the *number of residents* living in a particular territory, or the *volume of economic production* or other economic indicators.

Real use of ES can be expressed by indicators of *timber production* (volume of production) in a given period in the territory of different forest management areas. In case of timber outside of the forest, such statistics do not exist. The use of other types of fibres (special technical crops) can only be assessed through the use of crop areas and possibly the yield of these crops.

3.3 Drinking Water (P3)

3.3.1 Definition and Brief Characteristics of ES

Drinking water is one of the typical ecosystem products used every day. It is essential for life on Earth, for the development of society and for human well-being. Water intended for consumption is fully dependent on natural conditions or ecosystems. It is particularly at risk by anthropogenic influences (Fisher et al. 2009).

Freshwater ecosystems (rivers, lakes, coastal waters and groundwater) support the provision of more than one type of ES, such as water supply, fishing and recreation. Other important ES are also associated with the hydrological cycle in the watershed – e.g. water purification, water-flow regulation and climate regulation. The supply of drinking water is one of the provisioning ES, which are evident and in most cases economically valued, e.g. by suppliers of this commodity (Grizzetti et al. 2016).

Drinking water is one of the ecosystem services that are used directly. In our conditions and environment, the source of drinking water is mainly groundwater, to a lesser extent surface water (clean rainwater concentrated in rivers, reservoirs and lakes). Drinking water represents only part of the total surface and groundwater volume – only that part which meets certain quality criteria. In the SR these criteria are stipulated by the Act of the National Council of the SR no. 355/2007 Coll. on the protection, promotion and development of public health. Pursuant to Article 17 of this Act drinking water is:

> … in its original state or after treatment intended for drinking, cooking, preparation of food or other domestic purposes without regard to its origin and to whether it was delivered from a distribution network, water tank or as water packed into consumer packaging and water used in food factories in the production, treatment, conservation or sale of products or substances intended for human consumption.

EU legislation uses the equivalent term *water for human consumption* (available online: www.enviroportal.sk)

In EU countries, groundwater resources cover about 50% of drinking water consumption, surface water covers 37%, with the remaining 13% consisting of other anthropogenic sources (filtration, recycling) – (Fig. 3.10) (Schröter et al. 2019). In the SR, mainly underground sources (82%) are used for drinking water consumption, surface sources account for 18% of consumption (available online: www.pitnavoda.enviroportal.sk).

Overall, the ecosystem service of drinking water supply can be expressed as the *capacity of ecosystems (landscapes) to capture sufficient harmless water resources* that are or can be used for drinking purposes. Thus, in addition to the capacity to capture a certain amount of water resources, its quality aspect is also important – for this purpose, it is also necessary to achieve and maintain an adequate quality of the main environmental components (except for water, e.g. soil and substrate). The ultimate user of this ES is the population – unlike the ES Freshwater, which is used, besides the population, for various economic activities.

3.3.2 Methods Used to Assess and Identify ES

Freshwater ES are often quantified by biophysical and economic assessment methods which include standard science practices and models. Spatial models based on GIS and modelling such as QUICKScan, ESTIMAP, InVEST, State and Transition

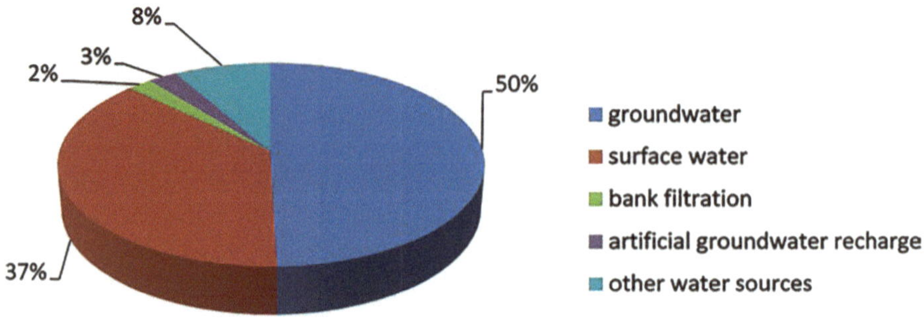

Fig. 3.10 The share of drinking water sources in the 27 EU countries. (Source: Schröter et al. 2019)

models, etc. are also very commonly used. Most of them rely on the spatial mapping of ecosystems and land use.

Biophysical basis for the provision of ES Drinking water is the presence of water reserves. According to Vačkář et al. (2014), the ES capacity is determined by the total amount of water (e.g. m^3/ha), while the performance indicator (sustainability of service) is the maximum sustainable water extraction (m^3/ha/year). However, not only water supplies are important – the amount and quality of water are influenced by complex climate interactions, topography, geology, land cover and management and other anthropogenic influences (Grizzetti et al. 2016). It is possible to obtain background data from the national statistics of the given country (e.g. from the Slovak Hydrometeorological Institute in Slovakia). Spatial projection and water reserves in territorial units have been taken into account in several countries (Denmark, Finland, Luxembourg, Romania) when assessing this ES. The rate of utilization of these reserves is often mentioned (pumped water with respect to reserves – Germany, Romania). More comprehensive indicators of the landscape's potential include, e.g. climatic parameters (precipitation balance), land use, soil and water quality, hydrogeology and vegetation quality (Pérez-Soba et al. 2015).

ES of drinking water can also be assessed from an economic point of view, with three available approaches – cost-based approaches, revealed preferences approaches or stated preferences approaches (Grizzetti et al. 2016). Several national ES assessments are based on the supply/demand formula for drinking water. Available national databases are used for the supply of drinking water as they include data on drinking water supplies, potential or production in the region. Demand for drinking water can be expressed as drinking water consumption, or its price, or as the number of residents living in a particular territory.

For example, the ES national assessment of Spain (Santos-Martín et al. 2016) calculates the contribution of surface water to drinking water, estimated on the basis of the European model. The surface water volume (m^3/ha/year) was expressed as the contribution of inland waterbodies and marshes to the drinking water supply. These values were recalculated according to the price of water, and the result presents the economic value of drinking water (€/ha/year). To determine the demand, water consumption statistics were used, which were subsequently converted to m^3/ha/year or to €/ha/year. Based on the data on drinking water supply and its consumption, an index was calculated, which represents the difference between supply and demand – in this way it is possible to determine the areas with a surplus or deficit of drinking water and express these values in biophysical and economic units. The national ES assessment in Germany (Rabe et al. 2016) is based on a similar principle but uses groundwater data. The input for the calculation is the capacity of the territory to produce drinking water based on soil infiltration capacity, zone saturation, information on groundwater drinking sources as well as factors that are not directly dependent on ecosystems. The definition of water consumption was based on data from all drinking water providers over a period of 3 years. From these inputs, the groundwater usage index was calculated, showing water usage as a percentage of the available quantity of reserves.

3.3.3 The Main Types of Landscape and Ecosystems Which Provide ES

Simply put, almost all types of ecosystems provide hydrological services, of course, to a different extent. Every ecosystem affects the properties of the water which runs through it (Brauman et al. 2007). Plants, as an important component of almost every ecosystem, act as natural filters which remove impurities and sediments before storing water in the recipient. In turn, forests and permanent vegetation affect how much water will be available in a given location and also affect water quality.

For drinking water production, the most important *ecosystem* is the geological environment in which groundwater is collected and from which most of the usable drinking water reserves are obtained (Fig. 3.11), which either comes to the surface via natural water springs or artificial wells and boreholes. From a global perspective, groundwater is particularly vulnerable to contamination of anthropogenic origin (chemicals, waste, nutrients, etc.), climate change and excessive groundwater abstraction. Each of these threats has the potential to alter the structure and functioning of groundwater and thereby provision of ES Drinking water, while these threats are concurrent (Schröter et al. 2019).

The amount of groundwater itself depends on their long-term reserves in different underground levels, partly also from immediate- and short-term processes and phenomena on the surface, in the subsurface layer of the soil and in layers of shallow groundwater circulation (Fig. 3.12). It is the subsurface layer that is heavily influenced by the amount of reserves, regime and quality characteristics as well as by the state and use of other types of ecosystems. In this respect, the most important types of ecosystems with the capacity to positively influence the provision of ES Drinking water include especially wetland ecosystems; forest ecosystems and permanent vegetation in the landscape; to some extent, grasslands and permanent agricultural crops.

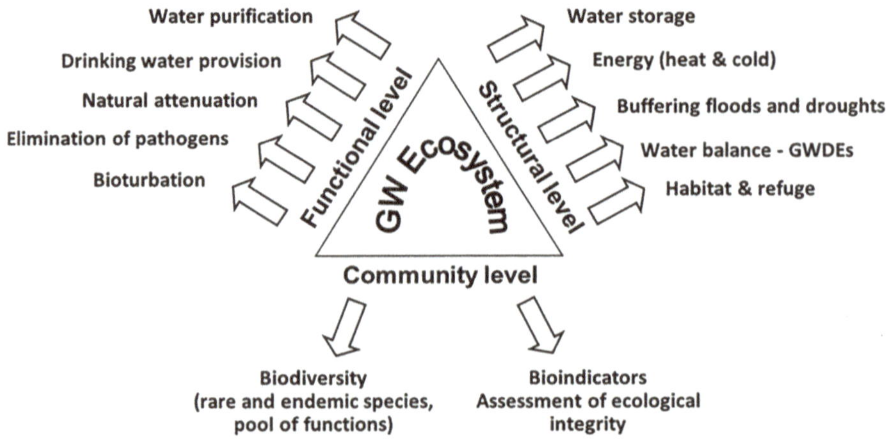

Fig. 3.11 Groundwater ecosystem services. (Source: Schröter et al. 2019)

Fig. 3.12 Springs and groundwater as an important part of ES Drinking water – Vojtovský prameň spring, Čadca. (Source: D. Kaisová)

The second most important source of drinking water comes from freshwater surface water resources (water reservoirs and watercourses). The threat to these resources comes mainly from the loss and destruction of habitats, river embankment and widespread water regime changes, environmental pollution and eutrophication, invasive species occurrence, climate change, and intensive use of water for other purposes (transport, energy, agriculture and fish stock). In terms of maintaining and improving the quality of surface water resources, the ecosystems similar to those for groundwater are the most important.

Water quality indicators are expressed by selected physical, chemical, biological and microbiological properties. In Slovakia, 28 water quality indicators are regularly monitored (e.g. presence of coliform bacteria and enterococci, microorganisms; content of selected metals, nitrates, nitrites and fluorides; biological and chemical oxygen consumption, water acidity, colour, turbidity and radiological indicators). It is the secondary contamination of water by human and animal faeces that is considered to be the greatest threat to drinking water – global mortality from drinking-water-related diseases exceeds 5 mil. people a year (Schröter et al. 2019).

3.3.4 Importance of ES in Terms of Nature and Landscape Protection in Slovakia

The hydrological component is one of the essential parts of ecosystems – it is essential for the existence and for the proper functioning of almost all types of ecosystems. ES Drinking water is especially important for humans – our nutrition and health – and to some extent for the existence and health of animals.

A large number of anthropogenic impacts on water resources have a significant influence on the quality of drinking water in Europe as well as in Slovakia. Drinking water, as one of the most important ES, is a vulnerable resource, protection of which requires integrated management. Extensive water purification is necessary, but it is not the only solution. Instead, the system's resilience should be strengthened by consistently reducing polluting factors. Responsible polluters should be actively involved in measures to reduce these threats, including the context of relevant legislation (Schröter et al. 2019).

In terms of nature and landscape protection, the proper functioning of the ES Freshwater is crucial. For this area (and for the favourable development of ecosystems and habitats in general), drinking water is not the most determining factor – of course, some quality status of water resources is particularly needed for animal nutrition. The relationship between water availability, hydrological regime on the one hand and nature and landscape protection on the other is described in the characteristic of ES Freshwater.

On the other hand, good practice of nature and landscape protection is a relatively important positive factor for better availability and quality of drinking water in adjacent areas (Fig. 3.13). Maintaining natural forests, non-intervention areas and appropriate management of protected areas contributes significantly to the mitigation of climate extremes, to the improvement of the hydrological regime of the area and thus to the increase of drinking water reserves in the landscape and to its better usability.

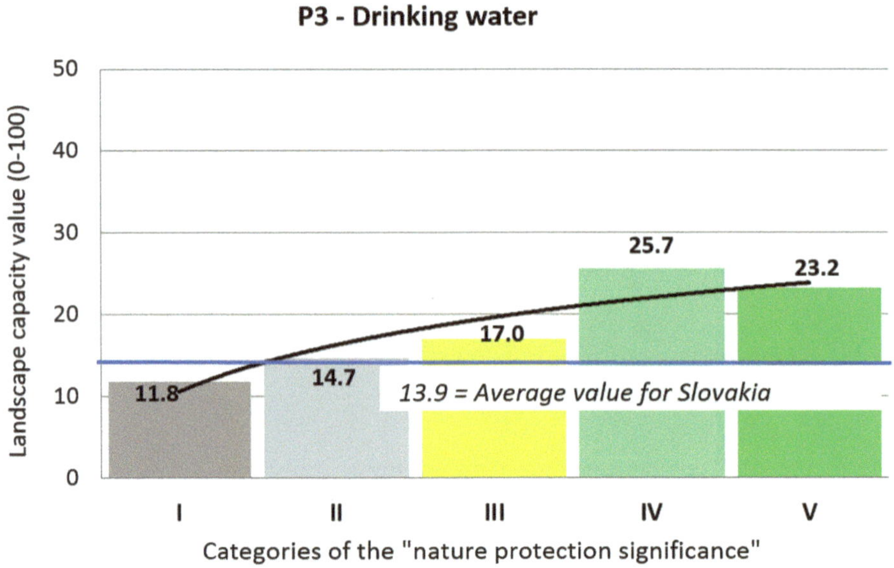

Fig. 3.13 The relationship between ecosystem service P3 and the significance of the territory of Slovakia in terms of nature and landscape protection

3.3.5 ES Assessment for the Territory of Slovakia

Drinking water sources in Slovakia are mostly located subsurface (springs, wells, boreholes) and to a lesser extent on the surface (water reservoirs and watercourses). The distribution of drinking water is provided via the public water supply network and to a lesser extent by individual wells. Recording and monitoring of water supplies (both drinking and freshwater) as well as its protection is ensured in long-term by organizations within the purview of the MoE (WRI, SWME).

At present, ten protected water management areas with a total area of 6942 km^2 have been declared in the Slovak Republic, representing 14.2% of the Slovak territory. About 1138 hygienic protection zones (HPZ) of groundwater sources are established in the SR. For the abstraction of surface water for drinking purposes, 73 HPZ have been set up in the territory of the Slovak Republic, of which 8 are related to abstraction from water reservoirs (Fig. 3.14), and 65 HPZ is set for direct abstraction from surface streams, which are situated mainly in the region of Eastern Slovakia (available online: www.mpsr.sk). Slovakia uses 102 watercourses and 8 water reservoirs (available online: www.shmu.sk). A substantial part of the water from surface water sources flows to Slovakia from neighbouring countries – Austria, the Czech Republic and Ukraine – and only a small part comes from our territory (14%). Groundwater sources are distributed unevenly depending on geological conditions. In Slovakia, natural groundwater reserves account for 146.7 m^3/s, of which 51.7% are usable (available online: www.mpsr.sk).

Fig. 3.14 Water reservoirs and their use for drinking water purposes – Starina water reservoir. (Source: M. Jančovič)

In 2017, 45.0 mil. m^3 of drinking water was taken from surface sources, which meant 19% of total surface water abstraction; and 7855 l/s (247.6 mil. m^3 per year) of subsurface resources, which represents up to 74% of total groundwater abstraction. Water losses in the SR pipeline network, however, amount to 25% (MoE SR 2017). At present, 88.9% of the population of Slovakia (4.84 mil.) is provided with a supply of drinking water from public water network, while part of the municipalities is supplied from local sources and garden wells.

The total water balance in the territory of Slovakia is positive – the share of total abstraction to the overall water supply volume is less than 10%. The long-term abstraction of water is decreasing considerably, but a gradual slight increase is currently expected. Surface and groundwater reserves are likely to decline due to changing climatic conditions.

Importantly, the quality of drinking water on the territory of Slovakia is high, with the proportion of unsatisfactory samples in long-term being at a level of less than 2–2.5%. In 2017 this was only 0.3%. In international statistics, the availability of safe drinking water for Slovakia is reported at 100%.

The most important reservoir of usable drinking water in Slovakia is Žitný ostrov area. With its area, it forms the largest inland *island* within Europe with an area of 1885 km^2 (Dušek and Velísková 2017). Almost one third of the total usable groundwater quantities in the SR is concentrated here, with approximately 10 billion m^3 located below the surface (available online: www.shmu.sk). Another important source of drinking water represents the Starina water reservoir in the north-eastern part of Slovakia, which has an area of 3.11 km^2 with a volume of 59.8 mil. m^3. The

Fig. 3.15 Capacity of the landscape to provide ES Drinking water (P3)

Table 3.3 Input data for assessment of ES Drinking water

Input data/ ES	P3 – Drinking water
Capacity	Important water management areas: watercourses, water reservoirs; water resources and their HPZ
	Protected water management areas and watercourse basins
	Data on drinking water supply, discharge of water resources
	Standardization of ecosystems/landscape types supporting ES Drinking water (CLS map reclassification)
	Environmental limit data – e.g. environmental quality (water pollution in terms of drinking water parameters) – not yet taken into account
Demand	Data on population distribution – urban settlements and densely populated areas
	Number of inhabitants in municipalities/districts – recalculated according to average water consumption
	Drinking water consumption by municipalities/districts (place of drinking water delivery)
Flow	Surface water and groundwater abstraction from individual river basins and waterbodies (for drinking purposes) – drinking water abstraction areas, the quantity of water collected

longest water-supply river sections are located on the Ondava and Topľa watercourses (available online: www.shmu.sk).

The result of the pilot assessment of the landscape's capacity to provide ES P3 is shown in Fig. 3.15. Multiple layers were entered into the calculation (Table 3.3) – locations of water resources, water reservoirs, natural balneo-therapeutical resources and their protection zones, watercourse basins and protected water management areas. The land use was subsequently incorporated into the calculation as a coefficient of improvement or deterioration of the given level of capacity (reclassification of the landscape structure). The processing of the calculations, their scaling and the graphical representation was similar to that of the other ES. The results illustrate the spatial distribution and relative significance of water resource protection categories. For the future, it is advisable to consider supplementing environmental quality indicators – especially surface and groundwater quality, but also soil and air quality, possibly with the occurrence of threats in terms of water quality (large landfills, other environmental risks).

Table 3.3 also shows some indicators which should be assessed for future expression of the level of demand for ES Drinking water as well as its actual use. Demand is given similarly as in the case of other production ES mainly by the number of inhabitants living in the territory, in combination with registered or expected water consumption. The flow of this ES is conditional on real use – abstraction of surface and groundwater for drinking purposes in river basins or administrative units. Here, it will be also necessary to combine data for different spatial units – natural (river basins) and socio-economic (municipalities, districts) in the subsequent comparison of capacity, demand and flow of the ES.

3.4 Freshwater (P4)

3.4.1 Definition and Brief Characteristics of ES

Human society uses water not only for drinking but also for various other purposes. Traditionally, freshwater is considered to be water used in industry and agriculture, the water necessary to produce hydroenergy, water for sanitation, or water needed for fishing purposes and genetic resources, etc. (available online: www.freshwater-tools.eu). The term also includes water used for waste management, transport or recreation. At the same time, water is an essential input to food and fibre production and is also used for many basic and complementary activities (Preston and Raudsepp-Hearne 2017). Ensuring reliable water supply is therefore of great importance for the functioning of the whole society (Becerra-Jurado et al. 2016).

Freshwater is obtained similarly to drinking water, mainly from subsurface and surface sources. However, its resources are much larger compared to drinking water, because they do not have to meet all the specified quality parameters. Although such parameters exist in some cases (e.g. water for recreation), they are considerably less stringent than for drinking water.

Annually, humanity uses 1000–1700 billion m^3 of water from subsurface and surface sources, while the estimated supply of these resources is 1100–4500 billion m^3/year (Hoekstra and Wiedmann 2014). While in 1990 the total specific drinking water demand was 425 litres per capita per day, it was only 164.96 litres per capita per day in 2016 (available online: www.enviroportal.sk).

Water supply is part of the so-called hydrological ES – Brauman et al. (2007) defined these as including benefits to people produced by the effects of terrestrial ecosystems on freshwater, with each service being determined by features such as quantity, quality, location and timing. ES related to water quality have also been described in detail by Keeler et al. (2012).

Accordingly, ES Freshwater can be defined as the *capacity of ecosystems and landscape to capture sufficient water resources which are or can be used for different purposes*. The water quality aspect of this ES is mostly not important. The end-users of this ES include various economic and non-production activities and also to a lesser extent the population.

3.4.2 Methods Used to Assess and Identify ES

The pilot study by Maes et al. (2016) analysed the freshwater-related ES by ecosystem typology, taking into account services provided by rivers, lakes, groundwater and wetlands. A different approach was taken by Brauman et al. (2007), who analysed individual relevant *hydrological services*. These are two basic approaches to ES assessment associated with drinking or freshwater. Both approaches deal with the integration of all services, with the first analysing all ecosystems and the other integrating all processes within the basin. ES relevant to water management are those related to aquatic ecosystems, water and soil interaction in different ecosystems such as forests, agricultural land, wetlands and other waterbodies.

There is a great demand for assessment tools which estimate the impact of landscape management on water supply services, such as irrigation, domestic consumption or hydroenergy production. Methods for assessing drinking and freshwater are often overlapping, with the assessments being applicable mostly for both ES using different statistical data. ES national assessments focus mainly on the assessment of drinking water, which may appear to be more important than freshwater. On the other hand, in case of freshwater, several uses can be dealt with, and more resource capacity can be considered.

As mentioned in the previous chapter, ES Drinking water and Freshwater are mainly assessed by two basic groups of methods – biophysical and economic. The amount of produced (available) water is an essential biophysical attribute in assessing water-related provisioning ES. This includes the volume of water available for drinking, agricultural purposes or the volume of flood discharge (Brauman et al. 2007) or the amount of water available for energy production. Many suitable or applied assessment indicators used by Pérez-Soba et al. (2015) are given in the characteristics of ES Drinking water.

The use of models presents a suitable way to assess water-related ES – e.g. Guswa et al. (2014) investigated the link between hydrological modelling and ES relevant for river basin management. The InVEST Water Yield model – the main purpose of which is to assess the energy obtained from water – is used as a tool for assessment of provisioning freshwater services, but its outputs are much wider. The model primarily seeks to answer the question of how changes in landscape use affect annual surface water reserves and thus affect water energy production. The model calculates the relative contribution of each partial area to the annual water supply and the value of this contribution in terms of energy production. Inputs to the model include the average rainfall, evapotranspiration, soil depth limit for root growth, water retention capacity, land use map, river catchment areas, landscape management information, ES demand and water energy price. It then provides the following outputs: current evapotranspiration (mm/year), water yield (supply) (mm/year), water demand (m^3/year) and energy available for hydropower plant (kW/year). It is a classic biophysical modelling approach (available online: www.naturalcapitalproject.org).

3.4.3 *The Main Types of Landscape and Ecosystems Which Provide ES*

Freshwater is essential for the functioning of all terrestrial ecosystems. The condition of aquatic ecosystems affects the provision of many ES necessary for the wellbeing of society (Jäppinen and Heliölä 2015). Water-related ES are derived directly from aquatic ecosystems, including rivers (Fig. 3.16), lakes, floodplains, wetlands and their adjacent riparian areas. In the case of provisioning ES such as water supply, it is desirable to increase its amount. On the contrary, in case of regulatory ES such as flood mitigation, it is beneficial to slow down and reduce the runoff.

Although the ecosystems themselves do not create water in any significant manner, but they do change the amount of water circulating in the landscape (Brauman et al. 2007) and thus influence the hydrological regime. This regime is dependent on a number of external factors which are influenced by the structure and status of ecosystems and the related ecosystem processes. The whole hydrological system is also greatly influenced by human activity, which is certainly one of the most important external factors influencing not only the quality but also the quantity and regime of water resources.

Fig. 3.16 Water in watercourses as an important source of freshwater – the Turiec river used to feed horses and cattle. (Source: J. Černecký)

ES Freshwater is directly provided by freshwater ecosystems, namely, surface and groundwater (of course, these include marine and ocean ecosystems in a wider European and global context). Water reservoirs, larger watercourses and groundwater sources are considered the most important ones. However, from the above mentioned, all wetland ecosystems, forest ecosystems and permanent vegetation in the landscape can also be included among (as in the case of ES Drinking water) other types of ecosystems which have a positive impact on the provision of this ES. These cocreate the *hydrological regime and the water cycle in the landscape* and contribute to the transformation of runoff, its attenuation and the creation of long-term reserves of surface and partly groundwater. To some extent, grasslands and permanent agricultural cultures can also be included in this category. On the other hand, urbanized areas, especially built-up areas and sealed surfaces, are considered to be the least contributing ecosystems to the provision of this ES.

3.4.4 Importance of ES in Terms of Nature and Landscape Protection in Slovakia

ES Freshwater is crucial not only for natural and seminatural ecosystems (which form a priority of nature and landscape protection) but also for the functioning of ecosystems used by people, which are again represented by various forms of land use. The availability of freshwater and the proper functioning of this ES can, therefore, be considered as one of the important preconditions for good health of habitats, which are important for nature and landscape protection in Slovakia. Areas with sufficient water reserves and an appropriate hydrological regime have proper conditions for the favourable state of protected areas of nature as well as conditions for the occurrence of protected and endangered species of plants and animals. Obviously, aquatic and wetland habitats are among the most important for the availability of water and a favourable hydrological regime. The same habitats are also included in the priorities with respect to nature and landscape protection.

It is the changes in land use, continuing urbanization and high water consumption during irrigation and climate change, which are currently changing the state of water reserves and the nature of the hydrological cycle. The results show, for example, that agriculture uses about 30% of the total amount of water in Europe and up to 80% of the total water collected in some parts of southern Europe. Currently, water scarcity affects about 10% of the European population, while 20–40% of water is lost unnecessarily in the pipeline network. Water consumption by the public, industry and agriculture is expected to increase by about 16% (Schröter et al. 2019), but as a result of climate change, the usable amount of water is likely to decrease. All this creates a great deal of pressure on the consumption of freshwater and drinking water and forms a precondition for the gradual reduction of its availability for natural ecosystems, protected areas, as well as plants and animals in the wild. Therefore, understanding the relationship between anthropogenic pressures

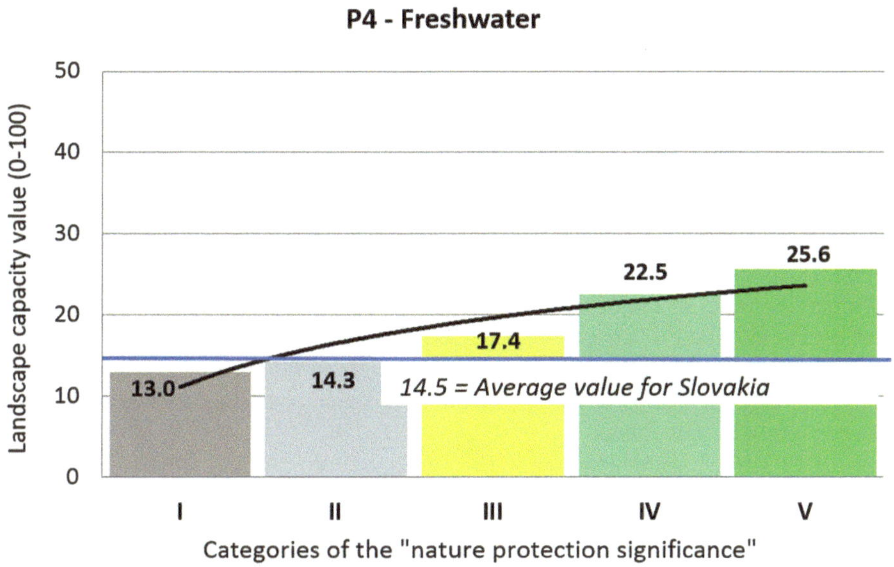

Fig. 3.17 The relationship between ecosystem service P4 and the significance of the territory of Slovakia in terms of nature and landscape protection

and the ecological status of aquatic ecosystems forms the basis for designing effective measures to achieve good ecological status not only for waterbodies and for water production but also for nature and landscape protection. In recent years, this fact has been also directly related to Slovakia, when a high water yield drop has been recorded, e.g. in the area of Záhorie, but also in many other areas.

On the other hand, good nature conservation practice supports the landscape's capacity to provide this ES. Natural and seminatural ecosystems, left to spontaneous development or appropriately managed, can contribute to improving the landscape's hydrological regime and thereby improve the availability and spatial distribution of freshwater within a landscape (see Fig. 3.17). These mainly include natural watercourses and areas, wetland communities (marshes, bogs, riparian and alluvial forests), wet meadows as well as natural forests of various types with varied species and age structures.

3.4.5 ES Assessment for the Territory of Slovakia

Based on hydrological assessment and surveys, there was 76,508 l/s of available groundwater volume in the SR in 2017 – in the long-term assessment, the increase in usable volume compared to 1990 is 1733 l/s, i.e. 2.3% (available online: www.eviroportal.sk). Slovakia is divided into 110 hydrogeological regions, and their balance status is mostly good. Of the total reserves, 10,607 l/s was being used, i.e.

13.9% of the total available volume (available online: www.shmu.sk). The freshwater from the total amount of groundwater collected represents 26% – mainly used in agriculture and in various other industries. In terms of spatial distribution, these available volumes of groundwater in Slovakia are diversified considerably.

A substantial part of the surface water fund of Slovakia flows from neighbouring countries, and its usability is limited. In total, about 2514 m^3/s of water flows in the long-term average, representing about 86% of the total surface reserves. Approximately 398 m^3/s of water originate from Slovakia considering the long-term average, accounting for 14% of the water fund and is thus not sufficient to meet the commercial needs of major economic and residential agglomerations. It is important that, in addition to the commercial requirements, the authorized quantities of surface water abstracted respect the requirements for the ecological limits of the relevant waterbody, so that the exploitation of these resources does not damage adjacent aquatic ecosystems.

In 2017, surface water abstraction amounted to 244.1 mil. m^3, which represented a decrease of 567.4 mil. m^3 compared to 1997 and 492.9 mil. m^3 compared to 2000. More than 80% of this volume was formed by freshwater, which is mainly used in industry and energy, to a lesser extent in agriculture (Fig. 3.18, available online: www.enviroportal.sk). According to the water management balance of water reservoirs for 2017, the total usable water volume is about 1300 mil. m^3. The total water reserves as of 1 January 2017 in the reservoirs (32, of which 20 are accumulation reservoirs) for 2016 amounted to 926.6 mil. m^3, which represents 80% of the total usable water volume (available online: www.shmu.sk).

In addition to the quantity of water resources, their quality is important. The overall quality of surface water in Slovakia is assessed in more than 1500 natural formations (watercourses and waterbodies). In the reference period from 2009 to

Fig. 3.18 Amelioration canals are significant water management elements, especially in the lowland country (Podunajská nížina). (Source: J. Špulerová)

2012, poor and very poor status was documented in about 9% of the number of monitored waterbodies, with 56% of the bodies showing very good and good ecological status. In terms of bathing water quality, a total of 79 natural waterbodies were assessed in 2017, with 25.4% of the samples being unsatisfactory. Groundwater quality is also regularly monitored in 75 bodies of groundwater (quaternary and pre-quaternary formations). In 2017, 14% of groundwater bodies were showing poor chemical status.

Pilot assessment of the landscape's capacity to provide ES P4 Freshwater was performed in a similar way to the ES Drinking water. Layers entering the calculation are listed in Table 3.4 – these include, in particular, groundwater reserves in individual regions, water reservoirs and watercourses significant from the water management point of view. The method of land use subsequently entered the calculation as a coefficient of improvement or worsening of the given level of capacity (reclassification of the landscape structure). The assessment results are shown in Fig. 3.19, in which the relative potential of the regions of Slovakia in terms of total surface and groundwater supply is expressed.

As regards the assessment of supply and demand with respect to ES Freshwater, similar indicators as those used for ES Drinking water may be considered suitable. Demand is determined by the intensity of use of freshwater resources – it is registered or assumed water consumption. It also depends to a certain extent on the number of inhabitants living in the territory, but the level of economic activity is also very important. The flow and real use of ES are given by surface and groundwater abstraction for commercial purposes in river basins or administrative units.

Table 3.4 Input data for assessment of ES Freshwater

Input data/ ES	P4 – Freshwater
Capacity	Hydrogeological regions + groundwater reserves
	Hydropedological data (permeability, infiltration ability)
	Watercourses significant for the water management
	Water reservoirs according to purpose
	Classification of ecosystems/landscape types supporting ES Freshwater (CLS map reclassification)
	Other suitable data:
	Watercourses with average discharge values
	Climatological and hydrological data – precipitation, evaporation, precipitation-runoff balance
Demand	Data on population distribution – urban settlements and densely populated areas
	Consumption of freshwater by municipalities/districts (water delivery areas)
	Freshwater requirements (municipalities, agricultural and industrial enterprises, etc.)
Flow	Quantities of surface water and groundwater abstraction from individual river basins and waterbodies – water delivery areas

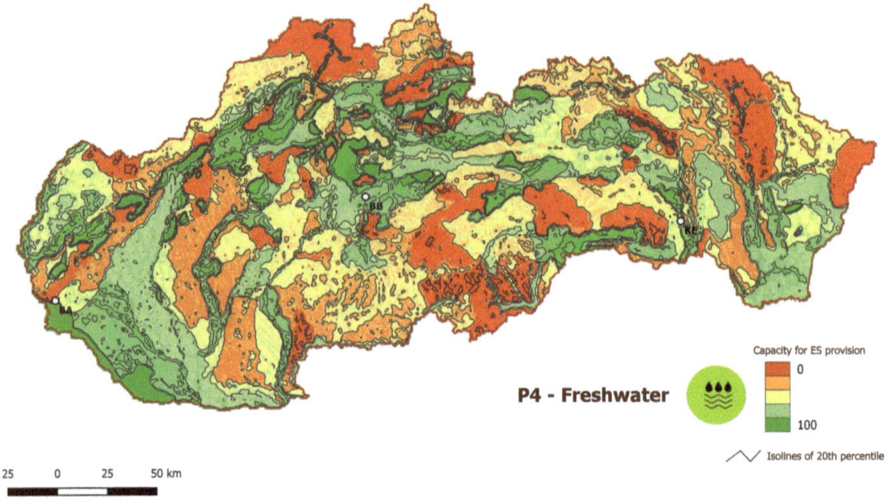

Fig. 3.19 Capacity of the landscape to provide ES Freshwater (P4)

3.5 Fish and Game/Wildfood (P5)

3.5.1 Definition and Brief Characteristics of ES

In addition to other functions, ecosystems provide people with fish, game and wildfood. According to Preston and Raudsepp-Hearne (2017), the ultimate benefit, in this case, comes in form of edible products derived from animals, plants and fungi which people use for their nutrition or commercial purposes (in addition to meet this includes fruits, seeds, mushrooms, tubers/roots, herbs, oils and the like).

Game hunting is considered an important part of rural traditions not only in Slovakia (Špiaková and Jančo 2017) but, for example, also in Sweden (Ljung et al. 2012) and Greece (Tsachalidis and Hadjisterkotis 2008). In Scotland and other parts of the UK, hunting is an old tradition which is important in defining social status (MacMillan and Leitch 2008). The economic factor also plays an important role in hunting. In countries with high average incomes, hunting is considered to be a symbol of the rich, with higher importance being attributed to the benefits of cultural ES (Murray and Simcox 2003), whereas in countries with lower average income, game hunting is primarily a provisioning service – a source of food (Tsachalidis and Hadjisterkotis 2008). The notion of wild game (from the point of view of provisioning ES in Slovakia) represents hunting game (including feathered game) and freshwater fish.

Maes et al. (2013) define the ES Wildfood as forest fruits (berries, fruits, nuts, mushrooms and seeds) and plants which are collected in nature to be used as food, or for medical purposes. The main incentives for collecting wildfood primarily include their availability and use in traditional cuisine (Pieroni 1999), medical purposes (Cruz Garcia and Price 2012) and their resale or artistic creations (Landor-Yamagata et al. 2018). In addition to the provisioning service, the collection of wildfood is also considered to be part of the identity of people, local traditions and the manifestation of natural and cultural heritage within cultural ES (Seeland and Staniszewski 2007). The collection is usually carried out in close proximity to human settlements (Stryamets et al. 2012), but higher participation rate is visible in rural communities (Łuczaj 2012).

The interest in collecting wildfood is decreasing due to population ageing and increasing urbanization. This occurs in many parts of Europe including Spain, Italy and Poland (Hadjichambis et al. 2008; Łuczaj 2010; Seeland and Staniszewski 2007). On the other hand, there is an assumption that the influence of industrialization and globalization will also cause an increased interest in the collection of wildfood, mainly due to cultural motives and traditions (Łuczaj 2012; Menendez-Baceta et al. 2012).

3.5.2 Methods Used to Assess and Identify ES

National assessment of ES Fish and Game/Wildfood was carried out in Finland (Jäppinen and Heliölä 2015), the Czech Republic (Vačkář et al. 2014), Romania (NEPA 2017), Luxembourg (Becerra-Jurado et al. 2016) and partly for selected protected areas also in Slovakia (Považan et al. 2015). Wildfood indicators have been proposed to assess total production based on the estimation of wildfood production per unit area. In Finland and Slovakia, the questionnaire survey method was used, with the national statistics (hunting licenses, hunting game total amount, hunted game amount, fishing permits) being the main source in Romania and Luxembourg. For most assessments of ES Fish and Game/Wildfood, a regional level was used. Input data included land cover type and land use in GIS, and also statistics on the amount of service produced (in kg/ha or number) were used to identify locations of supply and demand.

ES Fish and Game/Wildfood is not a commercially traded service but is a common part of market relations. The assessment method for calculating market prices proposed by Považan et al. (2014a) can be used for both wildfood and fish and game. Typical products which are included in the category of wildfood are medicinal plants, mushrooms, forest fruits and natural fibres. The value of wildfood V_{NT} is the price at which local people sell these products, e.g. to tourists or processing plants. The value of fishing and hunting V_{FH} mostly consists of two parts, namely, the value of the products sold, determined by market prices and the value of the license set by the national or local price level. These methods for calculating the value of wildfood, fish and game are described in more detail in the section of ES assessment for the territory of Slovakia.

Mapping the potential and occurrence of forest fruits and mushrooms and game hunting is also carried out by biophysical methods – e.g. in the environment of geographic information systems, it is possible to use a combination of landscape cover layers, habitat maps of assessed species with different spatial layers (e.g. climatic, hydrological factors, landscape management) and with data on the accessibility of the given territories as well as the time required for getting to the territory from inhabited settlements. Sampling-based direct mapping is appropriate for data acquisition in areas for which there is insufficient data (Burkhard and Maes 2017). It is possible to use statistical data from national databases on the population of individual species and data from hunting associations on the number of individual catches, the European Federation for Hunting and Conservation (FACE 2012).

The sociocultural assessment methods include in particular, questionnaires, semi-structured interviews and observations of gatherer, hunters or fishermen during their activities. These methods are suitable for detecting the species, quantity, area of harvest of wildfood and hunting of game and fishing. They also provide an opportunity for stakeholder representatives to express their views, which can be integrated into the decision-making process that is necessary for the proper management of the landscape (Chambers 2010; Sayer et al. 2013). Participative methods of mapping include mapping through analogue or digital data with participation from stakeholders and native people. In this way, we can identify the habitats of individual species of wildlife or forest fruits (Burkhard and Maes 2017).

3.5.3 The Main Types of Landscape and Ecosystems Which Provide ES

Wild game, such as roe deer or deer typically, occurs in forest communities (Fig. 3.20). Meadows, forest edges, fields as well as shrubs and hedges are suitable ecosystems for wild boar, pheasants or hares (Eliáš 2011). The landscape in which these species of game occurs should be as heterogeneous as possible to provide a shelter for the game. The agricultural landscape is, therefore, less suitable if it has large block fields without line vegetation and hedges where the game has no possibilities for shelter and breeding.

Watercourses are important for fish, such as unregulated rivers and streams, as well as waterbodies such as lakes (Jäppinen and Heliölä 2015). Lakes, ponds, watercourses and their riparian vegetation form typical environments where water birds live. They are being threatened by the removal of shrubs from the surroundings of watercourses, either for the purpose of flood control or biomass recovery.

Ecosystems, in which forest fruits grow, such as blueberries and cranberries, are represented by mountain grassland areas, or forest communities, where these plants form undergrowth. They occur mainly in areas with acidic soils. Raspberries and

Fig. 3.20 Mimicry of a young red deer (*Cervus elaphus*) merging with the environment of the forest ecosystem as part of an important ES for man in the form of a future hunting game. (Source: J. Černecký)

blackberries favour the clear-cuts or the edges of forest roads where they have enough sunlight to grow (Eliáš 1991). Forest communities are typical for mushroom growth, but some species can be found on meadows and pastures. The collection of medicinal plants is mainly linked to seminatural and natural grass-herb habitats, but pastures and forest communities are also important. Some species thrive on bare substrates (Eliáš 1991; Jäppinen and Heliölä 2015).

3.5.4 Importance of ES in Terms of Nature and Landscape Protection in Slovakia

Collecting wildfood, game hunting or fishing benefits both the gatherer/hunters and other consumers, whether as a source of food, income or cultural services – experiences. These services could positively influence nature and landscape protection management, through various obligations/measures related to the regulation of the overexploitation and displacement of natural species; or to the preservation of old forest stands as areas for breeding and conservation of game (Emanuelsson 2009). Nature and landscape protection is an essential part of preserving the provisioning service of collecting wildfood, game hunting and fishing. In most cases (whether it is wildfood collection, game hunting or fishing), their number is limited and bound to unique ecosystems that are very sensitive and responsive to change. At the same time, supporting nature and landscape protection in Slovakia promotes biodiversity. The more varied and valuable the territory, the better the prevalence of wildlife, precious plants and fruits, which can be described as a *positive correlation* relationship (see Fig. 3.21).

From the legislative point of view, the collection of freely growing wildfood in protected areas of Slovakia is regulated by Act no. 543/2002 Coll. on Nature and Landscape Protection, which states that the collection of plants, including their fruits, is prohibited for the general public in a territory where the third or higher level of protection applies. There are a number of reasons for such a strict limitation:

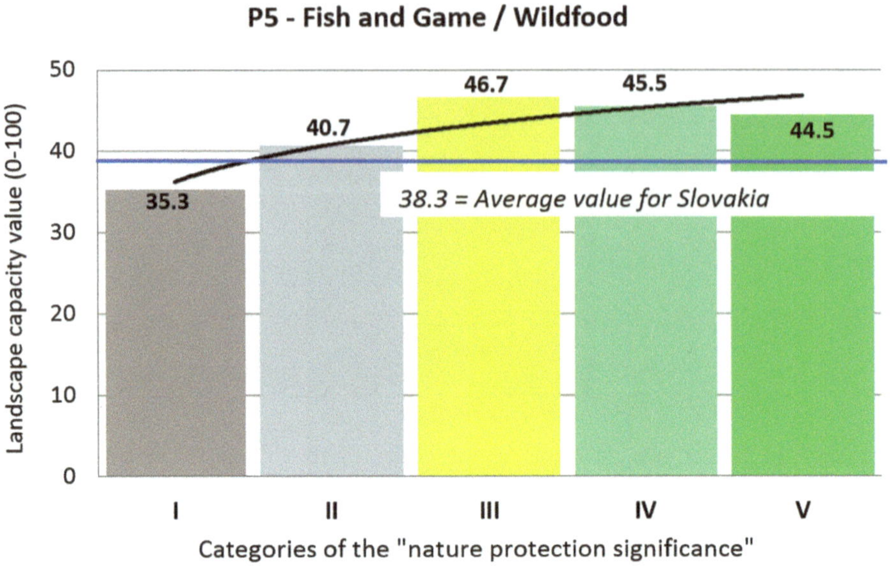

Fig. 3.21 The relationship between ecosystem service P5 and the significance of the territory of Slovakia in terms of nature and landscape protection

on the one hand, they are in many cases the only source of food for animals, of which there are a large number of protected ones (e.g. for western capercaillie (*Tetrao urogallus*) or brown bears (*Ursus arctos*), blueberries are significant seasonal food sources). On the other hand, in the case of insensitive collection of fruits with the help of various tools, the bushes and shrubs are damaged. Also, paths are created and these form a base for further erosion. Further reasons include disturbance of nature, waste and the pollution of the protected areas. The spatial restriction does not apply to protected species of plants and animals whose collection/hunting is prohibited throughout Slovakia. It is by collecting of plants that some protected species of European importance are endangered in the Slovak Republic, e.g. *Aconitum firmum* subsp. *moravicum* or *Daphne arbuscula* (Petrášová et al. 2013), as well as some animals – especially invertebrates (e.g. butterfly *Parnassius apollo*).

The hunting of game in Slovakia is governed by several Acts, especially Act no. 274/2009 Coll. on Hunting, as well as Act no. 543/2002 Coll. on nature and landscape protection, which limits the possibility of hunting game up to the fourth degree of nature protection (inclusive). The set of measures respecting nature protection also includes the Framework of Hunting Development of the SR 2017, as well as several regulations of the Ministry of Agriculture, directives, regulations and methodological instructions of the Slovak Hunting Union.

3.5.5 ES Assessment for the Territory of Slovakia

Formulas for calculating the value of wildfood and fish and game in the Carpathian protected areas are reported by Považan et al. (2014b). When calculating the value of wildfood, data from the management of the protected area, forest enterprises and national statistics may be used. The variables are as follows: yearly collection by product i – A_i (kg); average (median) product price at local level P_i (EUR/kg); i – type no. i. The value of nonwoody products: $V_{NT} = \Sigma A_i * P_i$. The data used comes from the reports of the relevant protected area administration, forest enterprises and national statistics (Považan et al. 2014b).

When calculating the value of wild game, it is appropriate to use data from reports of protected areas administrations, hunting associations, forest enterprises, national statistics and hunting statistical yearbook of the SR. The value of fishing and hunting V_{FH} mostly consists of two parts, namely, the value of the products sold, determined by market prices and the value of the license determined by the national or local price level. The variables are as follows: number of fishing and hunting licenses for I species – N_{FHi}; license price – P_{FHi} (EUR); number of units sold for species I – N_{Pi}; price of unit sold for species I – VPi. The value of fishing and hunting: $V_{FH} = \Sigma N_{FHi} * P_{FHi} + \Sigma N_{Pi} + V_{Pi}$ (Považan et al. 2014b).

For the spatial representation of the territories where the ES fish and game is located, it is possible to use the Forestry GIS, where the state administration maintains registers of hunting organizations, hunting grounds, users of hunting grounds and hunting guards (available online: www.forestportal.sk). Fishing grounds data can be retrieved for the fishing statistic.

An alternative way of assessment comes in the form of *value transfer* of the average cost of the game and wildfood, combined with the size of the ecosystems which produce the ES in some quality.

Due to the absence of statistical and other data, the assessment of the collection of wildfood (Fig. 3.22) in Slovakia is problematic. Considering the nature of the service, it is not possible to estimate the demand for this service (questionnaire survey) without more detailed research. The potential (supply) can be assessed based on data from the Catalogue of Habitats of Slovakia, the Atlas of Medicinal Plants and Berries and local case studies (Tutka et al. 2009; Považan et al. 2015).

For the assessment of ES Fish and Game/Wildfood for the territory of Slovakia in terms of capacity, demand and flow, it is important to obtain the correct data to be included in the assessment. The above-mentioned procedures and sources are suitable from the local to the regional level. That is why we have conducted the pilot assessment at the national level using appropriate and especially available spatial data (section Capacity in Table 3.5). The basic step was the reclassification of land use data (definition of CLS categories suitable for providing this service) and forestry (game enclosures, game reserves and pheasantries). At the basic level, hunting and fishing areas were defined according to the Landscape Atlas of the SR. These layers were the basis for determining the capacity of the landscape – data on the structure and quality of forest stands were then also included as the coefficients

Fig. 3.22 Collection of forest fruits as ES and also a part of the culture and traditions of Slovak people. (Source: J. Černecký)

Table 3.5 Input data for capacity, demand and flow assessment of ES Fish and Game/Wildfood

Input data/ES	P5 – Wildfood	Fish and Game
Capacity	Map of the current landscape structure – reclassification according to the suitability	Map of the current landscape structure – reclassification according to suitability for fish and game
	Forest stands quality and structure – reclassification	Game enclosures, game reserves and pheasantries – spatial projection
		Map of hunting/fishing grounds and areas
Demand	Data from questionnaire survey on demand for wildfood	Statistical data on the quantity of hunting permits/per unit area/per species
	Number of inhabitants and visitors of the municipality/region	Number of hunting/fishing permits
Flow	Statistical data on collected wildfood per unit area (kg/ha; kg/A)	Statistical data on the quantity of game hunted/fish caught per unit area (kg/ha; kg/A)
	Map of the current landscape structure	The real use of hunting/fishing grounds

improving or worsening this capacity. The overall assessment result is shown in Fig. 3.23.

Table 3.5 also shows the basic indicators which can be used for future expression of the level of ES demand as well as its real use. Demand can be expressed in a similar way to other provisioning ES by the number of inhabitants living in the territory, in combination with data on the number of hunting permits or by a survey among residents. The ES flow is conditioned by its real use – the amount of collected fruits, hunted game or caught fish. At national level, such data can only be obtained for regions, hunting associations or hunting grounds.

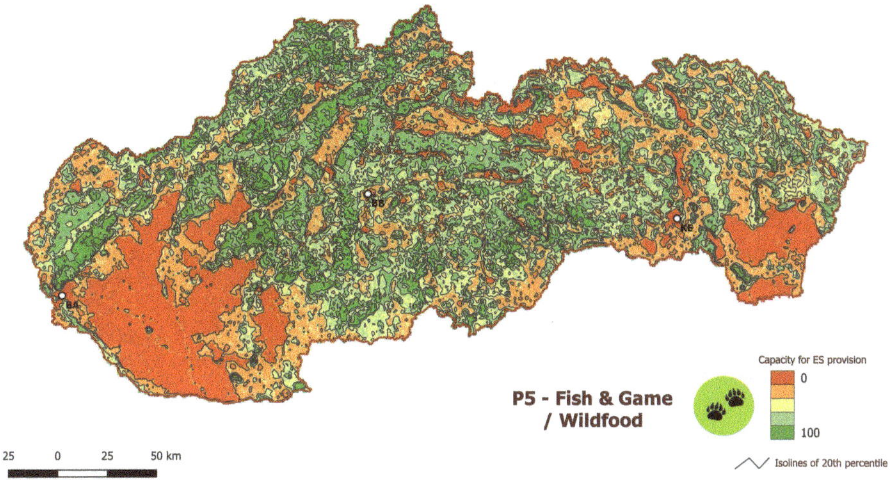

Fig. 3.23 Capacity of the landscape to provide ES Fish and Game/Wildfood

References

Becerra-Jurado, G., Philipsen, C., & Kleeschulte, S. (2016). *Mapping and assessing ecosystems and their services in Luxembourg – Assessment results*. Luxembourg: Le Gouvernment du Grand-Duché de Luxemburg. https://doi.org/10.13140/rg.2.1.4924.5841.

Bezáková, M. (2015). *Diverzita krajinných ekosystémov ako potenciál rozvoja vidieckeho regiónu v okrese Čadca*. Dissertation, Constantine the Philosopher University in Nitra.

Binder, S., Haight, R. G., Polasky, S., Warziniack, T., Mockrin, M., Deal, R., & Arthaud, G. (2017). *Assessment and valuation of forest ecosystem services: State of the science review*. Gen. Tech. Rep. NRS-170. Newtown Square, PA: U.S. Department of Agriculture, Forest Service, Northern Research Station.

Bradford, J. B., & D'amato, A. W. (2012). Recognizing trade-offs in multi-objective land management. *Frontiers in Ecology and the Environment, 10*(4), 210–216.

Brauman, K. A., Daily, G. C., Duarte, T. K., & Mooney, H. A. (2007). The nature and value of ecosystem services: An overview highlighting hydrologic services. *The Annual Review of Environment and Resources, 32*, 6.1–6.32.

Burkhard, B., & Maes, J. (Eds.). (2017). *Mapping ecosystem services. Bulgaria*. Sofia: Pensoft Publishers.

Carpentier, S., Elise, F., Tanya Handa, I., & Messier, C. (2017). Trade-offs between timber production, carbon stocking and habitat quality when managing timberlots for multiple ecosystem services. *Environmental Conservation, 44*(1), 14–23.

Chambers, R. (2010). *Paradigms, poverty and adaptive pluralism* (IDS Working Paper). Brighton, Institute of Development Studies Brighton.

Cruz Garcia, G. S., & Price, L. L. (2012). Weeds as important vegetables for farmers. *Acta Societatis Botanicorum Poloniae, 81*, 397–403. https://doi.org/10.5586/asbp.2012.047.

Czúcz, B., Arany, I., Potschin-Young, M., Bereczki, K., Kertész, M., Kiss, M., Aszalós, R., & Haines-Young, R. (2018). Where concepts meet the real world: A systematic review of ecosystem service indicators and their classification using CICES. *Ecosystem Services, 29*, 145–157. https://doi.org/10.1016/j.ecoser.2017.11.018.

De Meo, I., Cantiani, M., Ferretti, F., & Paletto, A. (2018). Qualitative assessment of forest ecosystem services: The stakeholders point of view in support of landscape planning. *Forests, 9*(8), 465. https://doi.org/10.3390/f9080465.

DEFRA. (2007). *An introductory guide to valuing ecosystem services*. London: Department for Environment, Food and Rural Affairs. https://www.gov.uk/government/publications/an-introductory-guide-to-valuing-ecosystem-services. Accessed 2 Mar 2019.

Duncker, P. S., Raulund-Rasmussen, K., Gundersen, P., et al. (2012). How forest management affects ecosystem services, including timber production and economic return: Synergies and trade-offs. *Ecology and Society, 17*(4), 50. https://doi.org/10.5751/ES-05066-170450.

Dušek, P., & Velísková, Y. (2017). Interaction between groundwater and surface water of channel network at Žitný Ostrov area. In M. A. Negm & M. Zeleňáková (Eds.), *Water Rresources in Slovakia: Part 1* (pp. 135–166). https://doi.org/10.1007/978-3-319-92853-1.

Džatko, M. (2002). *Hodnotenie produkčného potenciálu poľnohospodárskych pôd a pôdnoekologických regiónov Slovenska*. Bratislava: Research Institute of Soil Science and Soil Protection.

Eliáš, P. (1991). Prírodné zdroje liečivých rastlín na Slovensku, ich ochrana a racionálne využívanie. In *Pestovanie, zber a spracovanie liečivých rastlín: zborník referátov a posterov z medzinárodnej konferencie*, Slovakia, Vysoké Tatry 4–7 June 1991. Hlohovec: Slovakofarma.

Eliáš, P. (2011). Ekologické determinanty kvality života na vidieku: faktory zlepšujúce a faktory zhoršujúce kvalitu života ľudí. Paper presented at *Determinanty kvality života na vidieku: medzinárodná vedecká konferencie*, Slovak University of Agriculture, Nitra, 22–23 September 2011.

Emanuelsson, U. (2009). *The rural landscapes of Europe: How man has shaped European nature*. Stockholm: Formas.

European Commission. (2014). *Mapping and assessment of ecosystems and their services: Indicators for ecosystem assessments under Action 5 of the EU Biodiversity Strategy to 2020*. http://ec.europa.eu/environment/nature/knowledge/ecosystem_assessment/pdf/2ndMAESWorkingPaper.pdf. Accessed 2 Mar 2019.

FACE. (2012). *European Federation of Associations for Hunting and Conservation*. https://www.face.eu. Accessed 2 Apr 2019.

Fisher, B., Turner, R. K., & Morling, P. (2009). Defining and classifying ecosystem services for decision making. *Ecological Economics, 68*(3), 643–653. https://doi.org/10.1016/j.ecolecon.2008.09.014.

Grizzetti, B., Lanzanova, D., Liquete, C., Reynaud, A., Cardoso, A., & C. (2016). Assessing water ecosystem services for water resource management. *Environmental Science & Policy, 61*, 194–203. https://doi.org/10.1016/j.envsci.2016.04.008.

Guswa, A. J., et al. (2014). Ecosystem services: Challenges and opportunities for hydrologic modeling to support decision making. *Water Resources Research, 50*(5), 4535–4544. https://doi.org/10.1002/2014WR015497.

Hadjichambis, A. C. H., Paraskeva-Hadjichambi, D., Della, A., Giusti, M. E., De Pasquale, C., Lenzarini, C., Censorii, E., Gonzales-Tejero, M. R., Sanchez-Rojas, C. P., & Ramiro-Gutierrez, J. M. (2008). Wild and semi-domesticated food plant consumption in seven circum-Mediterranean areas. *International Journal of Food Sciences and Nutrition, 59*(5), 383–414. https://doi.org/10.1080/09637480701566495.

Hodgson, E. (2012). Human environments: Definition, scope, and the role of toxicology. *Progress in molecular biology and translational science. Academic Press, 112*, 1–10. https://doi.org/10.1016/B978-0-12-415813-9.00001-5.

Hoekstra, A. Y., & Wiedmann, T. O. (2014). Humanity's unsustainable environmental footprint. *Science, 344*(6188), 1114–1117. https://doi.org/10.1126/science.1248365.

Jäppinen, J. P., & Heliölä, J. (Eds.) (2015). *Towards a sustainable and genuinely green economy. The value and social significance of ecosystem Services in Finland (TEEB for Finland)*. Synthesis and Roadmap. Helsinki: The Finnish Ministry of Environment.

Kandziora, M., Burkhard, B., & Müller, F. (2013). Interactions of ecosystem properties, ecosystem integrity and ecosystem service indicators—A theoretical matrix exercise. *Ecological Indicators, 25*, 54–78. https://doi.org/10.1016/j.ecolind.2012.09.006.

Karuppuchamy, P., & Venugopal, S. (2016). Integrated pest management. https://doi.org/10.1016/B978-0-12-803265-7.00021-X.

Keeler, B. L., Polasky, S., Brauman, K. A., Johnson, K. A., Finlay, J. C., O'Neill, A., Kovacs, K., & Dalzell, B. (2012). Linking water quality and well-being for improved assessment and valuation of ecosystem services. *Proceedings of the National Academy of Sciences, 109*(45), 18619–18624. https://doi.org/10.1073/pnas.1215991109.

Landor-Yamagata, J., Kowarik, I., & Fischer, L. K. (2018). Urban foraging in Berlin: People, plants and practices within the metropolitan green infrastructure. *Sustainability, 10*(6), 1873. https://doi.org/10.3390/su10061873.

Ljung, P. E., Riley, S. J., Heberlein, T. A., & Ericsson, G. (2012). Eat prey and love: Game-meat consumption and attitudes toward hunting. *Wildlife Society Bulletin, 36*(4), 669–675. https://doi.org/10.1002/wsb.208.

Łuczaj, Ł. J. (2010). Changes in the utilization of wild green vegetables in Poland since the 19th century: A comparison of four ethnobotanical surveys. *Journal of Ethnopharmacology, 128*(2), 395–404. https://doi.org/10.1016/j.jep.2010.01.038.

Łuczaj, Ł. (2012) Ethnobotanical review of wild edible plants of Slovakia. *Acta Societatis Botanicorum Poloniae, 81*(4), 245–255.

Macmillan, D. C., & Leitch, K. (2008). Conservation with a gun: Understanding landowner attitudes to deer hunting in the Scottish highlands. *Human Ecology, 36*(4), 473–484. https://doi.org/10.1007/s10745-008-9170-9.

Maes, J., Teller, A., Erhard, M., Liquete, C., Braat, L., Berry, P., Egoh, B., Puydarrieux, P., Fiorina, C., Santos, F., Paracchini, M. L., Keune, H., Wittmer, H., Hauck, J., Fiala, I., Verburg, P. H., Condé, S., Schägner, J. P., San Miguel, J., Estregui, L. C., Ostermann, O., Barredo, J. I., Pereira, H. M., Stott, A., Laporte, V., Meiner, A., Olah, B., Royo Gelabert, E., Spyropoulou, R., Petersen, J. E., Maguire, C., Zal, N., Achilleo, S. E., Rubin, A., Ledoux, L., Brown, C., Raes, C., Jacobs, S., Vandewalle, M., Connor, D., & Bidoglio, G. (2013). *Mapping and assessment of ecosystems and their services. An analytical framework for ecosystem assessments under action 5 of the EU biodiversity strategy to 2020*. Luxembourg: Publications office of the European Union.

Maes, J., Teller, A., Erhard, M., Murphy, P., Paracchini, M. L., Barredo, J. I., Grizzetti, I. B., Cardoso, A., Somma, F., Petersen, J.-E., Meine, R. A., Gelabert, E. R., Zal, N., Kristensen, P., Bastrup-Birk, A., Biala, K., Romao, C., Piroddi, C. H., Egoh, B., Fiorina, C. H., Santo, F., Naruševičius, V., Verboven, J., Pereira, H., Bengtsson, J., Gocheva, K., Marta-Pedroso, C., Snäll, T., Estreguil, C. H., San Miguel, J., Braat, L., Grêt-Regamey, A., Perez-Soba, M., Degeorges, P., Beaufaron, G., Lillebø, A., Abdul Malak, D., Liquete, C., Condé, S., Moen, J., Östergård, H., Czúcz, B., Drakou, E. G., Zulian, G., & Lavalle, C. (2014). *Mapping and Assessment of Ecosystems and their Services (MAES). Indicators for ecosystem assessments under Action 5 of the EU Biodiversity Strategy to 2020 2nd Report – Final*. European Union.

Maes, J., Liguete, C., et al. (2016). An indicator framework for assessing ecosystem services in support of the EU biodiversity strategy to 2020. *Ecosystem Services, 17*, 14–23. https://doi.org/10.1016/j.ecoser.2015.10.023.

MEA – Millennium Ecosystem Assessment. (2005). *Ecosystems and human well-being: Wetlands and water synthesis*. Washington, DC: World Resources Institute.

Menendez-Baceta, G., Aceituno-Mata, L., Tardío, J., Reyes-García, V., & Pardo-De-Santayana, M. (2012). Wild edible plants traditionally gathered in Gorbeialdea (Biscay, Basque Country). *Genetic Resources and Crop Evolution, 59*(7), 1329–1347. https://doi.org/10.1007/s10722-011-9760-z.

MoE SR. (2017). *Envirorezort sa pripojil k iniciatíve za záchranu opeľovačov – Ministerstvo životného prostredia Slovenskej republiky*. http://www.minzp.sk/tlacovy-servis/tlacove-spravy/

tlacove-spravy-2017/tlacove-spravy-marec-2017/envirorezort-pripojil-k-iniciative-za-zachranu-opelovacov.html. Accessed 6 April 2019.

MPSR NLC. (2018). *Správa o lesnom hospodárstve v Slovenskej republike za rok 2017 – Zelená správa*. http://www.mpsr.sk/index.php?navID=123&id=13656. Accessed 4 Apr 2019.

Murray, M., & Simcox, H. (2003). Use of wild living resources in the United Kingdom – A review. In *UK committee for IUCN* (pp. 101–109). UK: IUCN.

NEPA. (2017). *Assessment of ecosystems and ecosystem services in Romania*. Romania: NEPA, NINA, ROSA, WWF Romania.

Pazúr, R., & Bolliger, J. (2017). Land changes in Slovakia: Past processes and future directions. *Applied Geography, 85*, 163–175.

Pérez-Soba, M., Harrison, P.A., Smith, A.C., Simpson, G., Uiterwijk, M., Miguel Ayala, L, Archaux, F., Erős, T., Fabrega Domenech, N., György, Á.I., Haines-Young, R., Li, S., Lommelen, E., Meiresonne, L., Mononen, L., Stange, E., Turkelboom, F., Veerkamp, C., & Wyllie De Echeverria, V. (2015). *Database and operational classification system of ecosystem service – Natural capital relationships. EU FP7 OpenNESS Project Deliverable 3.1*. European Commission. http://www.openness-project.eu/sites/default/files/OpenNESS_D3.1_Final.pdf. Accessed 6 April 2019.

Petrášová, A., Uhliarová, E., Sabo, P., & Považan, R. (2013). *Manažment chránených druhov rastlín*. Banská Bystrica: Belianum.

Pieroni, A. (1999). Gathered wild food plants in the upper valley of the Serchio River (Garfagnana), Central Italy. *Economic Botany, 53*, 327–341. https://doi.org/10.1007/BF02866645.

Považan, R., Getzner, M., & Švajda, J. (2014a). Value of ecosystem services in mountain national park. Case study of Vel'ká Fatra National Park (Slovakia). *Polish Journal of Environmental Studies, 23*(5), 1699–1710.

Považan, R., Getzner, M., & Kadlečík, J. (2014b). Hodnotenie ekosystémových služieb v chránených územiach Karpát so zameraním na Slovensko – Príručka pre rýchle hodnotenie. *Quaestiones rerum naturalium, 1*(2), 7–44.

Považan, R., Getzner, M., & Švajda, J. (2015). On the valuation of ecosystem services in Muránska Planina National Park (Slovakia). *Eco. Mont. Journal on Protected Mountain Areas Research and Management, 7*(2), 61–69. https://doi.org/10.1553/eco.mont-7-2s61.

Power, A. G. (2010). Ecosystem services and agriculture: Tradeoffs and synergies. *Philosophical Transactions of the Royal Society, B: Biological Sciences, 365*(1554), 2959–2971. https://doi.org/10.1098/rstb.2010.0143.

Preston, S. M., & Raudsepp-Hearne, C. (2017). *Ecosystem service toolkit: Completing and using ecosystem service assessment for decision-making: An interdisciplinary toolkit for managers and analysts* (Federal, Provincial, and Territorial Governments of Canada, Ottawa). Ottawa: Environment and Climate Change Canada Enquiry Centre.

Rabe, S.-E., Koellner, T., Marzelli, S., Schumacher, P., & Gret-Regamey, A. (2016). National ecosystem services mapping at multiple scales – The German exemplar. *Ecological Indicators, 70*, 357–372. https://doi.org/10.1016/j.ecolind.2016.05.043.

Santos-Martín, F., García Llorente, M., Quintas-Soriano, C., Zorrilla-Miras, P., Martín-López, B., Loureiro, M., Benayas, J., & Montes, M. (2016). *Spanish National Ecosystem Assessment: Socio-economic valuation of ecosystem services in Spain. Synthesis of the key findings*. Madrid: Biodiversity Foundation of the Spanish Ministry of Agriculture, Food and Environment.

Sayer, J., Sunderland, T., Ghazoul, J., Pfund, J. L., Sheil, D., Meijaard, E., Venter, M., Boedhihartono, A. K., Day, M., Garcia, C., Van Oosten, C., & Buck, L. E. (2013). Ten principles for a landscape approach to reconciling agriculture, conservation, and other competing land uses. *Proceedings of the National Academy of Sciences, 110*(21), 8349–8356. https://doi.org/10.1073/pnas.1210595110.

Schröter, M., Bonn, A., Klotz, S., Seppelt, R., & Baessler, C. (2019). *Atlas of ecosystem services. Drivers, risks, and societal responses*. Springer.

Seeland, K., & Staniszewski, P. (2007). Indicators for a European cross-country state-of-the-art. Assessment of non-timber forest products and services. *Small-scale Forestry, 6*, 411–422. https://doi.org/10.1007/s11842-007-9029-8.

Špiaková, J., & Jančo, M. (2017). Cledsimate and recreational hunting, a case study from the Liptov region, Slovakia. In J. Fialková & D. Pernicová (Eds.), *Public recreation and landscape protection – With nature hand in hand?* Brno

Stryamets, N., Elbakidze, M., & Angelstam, P. (2012). Role of non-timber forest products for local livelihoods in countries with transition and market economies: Case studies in Ukraine and Sweden. *Scandinavian Journal of Forest Research, 27*(1), 74–87. https://doi.org/10.1080/02827581.2011.629622.

TEEB – The Economics of Ecosystems and Biodiversity. (2013). *Guidance Manual for TEEB Country Studies*. Version 1.0.

Tsachalidis, E. P., & Hadjisterkotis, E. (2008). Wild boar hunting and socioeconomic trends in Northern Greece, 1993–2002. *European Journal of Wildlife Research, 54*(4), 643–649. https://doi.org/10.1007/s10344-008-0190-y.

Tutka, J., Vilček, J., & Kovalčík, M. (2009). *Oceňovanie verejnoprospešných funkcií lesných a poľnohospodárskych ekosystémov a služieb odvetví. In Abstracts of Aktuálne otázky ekonomiky lesného hospodárstva Slovenskej republiky*, Technical University in Zvolen, 2009.

UN SEEA. (2014). *System of environmental economic accounting 2012 – Central framework*. New York: United Nations.

Vačkář, D., Frélichová, J., Lorencová, E., Pártl, A., Harmáčková, Z., & Loučková, B. (2014). *Metodologický rámec integrovaného hodnocení ekosystémových služeb v České republice*. http://www.ecosystemservices.cz/cs/metodologicky-ramec-integrovaneho-hodnoceni-ekosystemovych-sluzeb-v-ceske-republice. Accessed 13 April 2019.

Vihervaara, P., Mononen, L., Nedkov S., Viinikka, A., et al. (2018). *Biophysical mapping and assessment methods for ecosystem services*. ESMERALDA Deliverable D3.3. http://www.esmeralda-project.eu/documents/1/. Accessed 10 Apr 2019.

Vilček, J. (2011). Potenciály a parametre kvality pol'nohospodárskych pôd Slovenska. *Geografický časopis, 63*(2), 133–154.

Vilček, J., & Koco, Š. (2018). Integrated index of agricultural soil qusality in Slovakia. *Journal of Maps, 14*(2), 68–76. https://doi.org/10.1080/17445647.2018.1428233.

Chapter 4
Regulatory Ecosystem Services and Supporting Ecosystem Functions

Ján Černecký, Jana Špulerová, Viktória Ďuricová, Peter Mederly, Martin Jančovič, Juraj Hreško, and Matej Močko

Abstract This chapter provides the analysis and assessment of **ten regulatory and supporting ES**: *R1, air quality regulation; R2, water quality regulation; R3, erosion and natural hazard regulation; R4, water flow regulation; R5, local climate regulation; R6, global climate regulation/carbon sequestration; R7, biodiversity promotion; R8, life cycle maintenance/pollination; R9, pest and diseases control; and R10, maintenance of soil formation and composition.* All ES are described in the unified structure: definition and brief characteristics, methods used for identification and assessment, main types of landscape and ecosystems providing given ES, the importance of ES in terms of nature and landscape protection, and ES assessment for the territory of Slovakia. Spatial assessment is provided as a map of the landscape capacity for given ES provision. For all ES, short conclusions and overview of input data for further assessment of the ES capacity, demand and flow are also given.

J. Černecký (✉)
Faculty of Natural Sciences, Constantine the Philosopher University in Nitra, Nitra, Slovakia

State Nature Conservancy of the Slovak Republic, Banská Bystrica, Slovakia

Institute of Landscape Ecology of the Slovak Academy of Sciences, Bratislava, Slovakia
e-mail: jan.cernecky@sopsr.sk

J. Špulerová
Institute of Landscape Ecology of the Slovak Academy of Sciences, Bratislava, Slovakia
e-mail: jana.spulerova@savba.sk

V. Ďuricová
State Nature Conservancy of the Slovak Republic, Banská Bystrica, Slovakia

Faculty of Natural Sciences, Matej Bel University, Banská Bystrica, Slovakia
e-mail: viktoria.duricova@sopsr.sk

P. Mederly · M. Jančovič · J. Hreško · M. Močko
Faculty of Natural Sciences, Constantine the Philosopher University in Nitra, Nitra, Slovakia
e-mail: pmederly@ukf.sk; martin.jancovic@ukf.sk; jhresko@ukf.sk; matej.mocko@ukf.sk

P. Mederly, J. Černecký (eds.), *A Catalogue of Ecosystem Services in Slovakia*,
https://doi.org/10.1007/978-3-030-46508-7_4

4.1 Air Quality Regulation (R1)

4.1.1 Definition and Brief Characteristics of ES

Burkhard and Maes (2017) identify *air pollution as one of the main environmental risks*, especially for urban areas, because of the high production and concentration of pollutants in the air. The source of air pollution comes mainly from anthropogenic activities and anthropogenic controlled ecosystems, which release pollutants into the atmosphere, and these can then be deposited elsewhere, also in pollution-sensitive ecosystems. For example, NH_3 and NO_2 emissions from livestock farming and the use of fertilizers (also used for ecosystem management) can lead to increased nitrogen deposition or direct intoxication of plants sensitive to this type of pollution (Sutton et al. 2011). Deposition of pollutants from the atmosphere in the soil and vegetation can significantly reduce their concentration in the air (Fowler et al. 2009) and thus reduce the adverse effects on human health and other ES (RoTAP 2012).

According to the UK National Ecosystem Assessment UK NEA, air quality regulation is a primary or intermediary regulatory service which affects atmospheric concentrations of air pollutants and their deposition in land and water. At the national level, the most important pollutants include particular matter, ozone, nitrogen oxides, ammonia and sulphur, deposition of which can lead to acidification of ecosystems and their eutrophication.

Ecosystems contribute to improving air quality by removing pollutants from the atmosphere: gases and solid particles are deposited on the ecosystem (especially plant) surfaces, and polluting gases enter the leaves through stomata. The extent of this removal depends on a number of factors, including air turbulence (higher vegetation has higher effectiveness), duration of foliage (evergreen trees are more effective) and stomatal processes (deposition may decrease under dry conditions – UKNEA 2011b).

Maintaining good air quality depends on the exchange of chemicals between ecosystems and the atmosphere through biogeochemical cycles. Soil, along with vegetation, emits compounds which contribute to the formation of secondary pollutants in the atmosphere, such as the emission of volatile organic carbon from plants, which contributes to the formation of ozone and aerosols in the ground layer of the atmosphere (Royal Society 2008).

Air quality regulation through ecosystems brings many benefits, including clean air for breathing, prevention of respiratory and skin diseases. Ecosystems affect air quality by emitting chemicals into the atmosphere (serving as a *source*) or extracting chemicals from the atmosphere – i.e. serve as *waste containers* for industrial emissions, for example sulphur compounds (Preston and Raudsepp-Hearne 2017). The removal of pollutants from the air is mainly performed by trees and other vegetation through dry deposition of substances which accumulate at the earth's surface (Burkhard and Maes 2017). The Spanish National Ecosystem Assessment (Santos-Martín et al. 2016) lists air quality regulation and climate regulation (57% of answers) as the most valuable benefits provided by ecosystems to maintain quality of life.

To conclude and summarize, *air quality regulation is the ES which mainly consists of attenuation/transformation of the effects of air pollution on ecosystems and people.*

4.1.2 Methods Used to Assess and Identify ES

Considering the physical-chemical nature of the processes associated with this ES, biophysical methods are mainly used for its assessment. The atmospheric gas flow, atmospheric/air-purifying capacity and pollutant level/content in the atmosphere are appropriate indicators for measuring air quality regulation. Burkhard and Maes (2017) present secondary (supporting) indicators important for air quality regulation: net primary production, disease prevention, regulation of ecosystem dynamics and stability of ecosystem processes, ability of ecosystem restoration, ecosystem diversity and interconnection promotion.

The mapping of ES air quality regulation according to Burkhard and Maes (2017) is based on three types of information: dry deposition rate (potential), air pollutant removal (real production) and human pollution exposure (demand). A good measure of this ES comes in the form of the cycle of pollutant removal through vegetation as a result of dry deposition and pollutant concentration. Consumption of this ES can be mapped based on population exposure and pollutant concentration above the limit set by legislation.

The Finnish national ES assessment (Jäpinen and Heliölä 2015) used a cascade model with four indicators: structure, function, benefit and value of ecosystem service provision. The following indicators are used in case of air quality regulation: green infrastructure in cities (structure), storage/absorption of small particles (function), improvement of air quality (benefit), health benefits from clean air and saved/avoided healthcare costs (value).

For the regulation of local climate and air quality, the national ES assessment in Germany (Albert et al. 2016) selected the indicators of the *extent of green areas in settlements* as the potential of ES provision. Germany has extensive environmental data available and considers the ES potential through assessing and planning at the regional and municipal level as part of landscape planning.

Modelling tools InVEST or ARIES need to be highlighted as comprehensive tools for ES assessment. Both models work in ArcGIS environment and are freely available. The primary input for these models is the land cover and land use maps, complemented by socio-economic and ecological parameters (carbon stock in soils, average annual rainfall). On the other hand, a simplified production matrix method is also used (Burkhard et al. 2014). Accordingly, air quality regulation is provided to the highest extent by forest ecosystems with an index of 5, while cities and densely populated areas have the highest consumption with an index of −5.

Another group of assessment methods is the economic methods of air quality regulation – according to Farber et al. (2006), contingent valuation method, cost savings or replacement costs methods could be used. For cost savings, the ES is valued on the basis of an estimate of costs which have not been incurred or the possibility of avoiding the costs associated with averting or mitigating the negative effects of the absence of the ES. The replacement cost method assesses the ES according to the cost of replacing this service.

The integrated ES assessment in the Czech Republic (CZ) is based on the assessment of the current ES status, including the regulation of air pollution from the *value transfer* method. In this way, the economic valuations of the given ES in many studies performed under comparable conditions were used, and the values were *transferred* in a new context in conditions of the CZ (Vačkář et al. 2014). The average economic value of ES air quality regulation by Frélichová et al. (2014) in the CZ is 266.33 EUR/ha.

4.1.3 The Main Types of Landscape and Ecosystems Which Provide ES

The national assessments of ecosystems and their services, together with the production matrix (Burkhard et al. 2014), confirm that the most widespread and important biotope on the European level providing ES air quality regulation is the forest ecosystems and another wooded land.

This also applies to Slovakia, where forest ecosystems are clearly the most important for air quality regulation. Other ecosystems are essentially of little significance from a nationwide perspective, but they can be significant locally. In the built-up areas, there is clearly the largest demand for this ES combined with its highest consumption. Forest ecosystems are therefore crucial both in terms of the quality of provision of this ES (Fig. 4.1) and in terms of the overall ecosystem area in Slovakia. It is important that all areas of Slovakia have a sufficient share of continuous forest stands, which is the case, in particular, in Central Slovakia. The southern parts of Western and Wastern Slovakia, which are dominated by arable land, are significantly poorer for the provision of this ES. It is essential to maintain/expand/restore urban parks and vegetation in cities, especially from a local point of view, so that these areas are as close as possible to the place of demand and consumption.

The area of forest stands in 2017 amounted to 1.9 mil. ha, i.e. more than 40% (38% based on the ecosystem map of Slovakia by Černecký et al. 2020) of the area of Slovakia (MPSR NLC 2018). Thanks to this fact, the forested landscape has the highest share in provision the air quality regulation. Among other functions, forest ecosystems play a key role in the deposition of pollutants from the air, and therefore their protection is crucial.

Fig. 4.1 Natural oak forest in SAC Mäsiarsky bok with old trees contributes significantly to air quality regulation. (Author: J. Černecký)

Ecosystems of good quality have a clearly positive effect on air quality, primarily through the absorption, storage and removal of pollutants. However, if the pollutant storage rate exceeds the critical thresholds, the opposite effect on the other ES may occur. Emissions in the atmosphere from ecosystems can even directly or indirectly deteriorate air quality (UK NEA).

4.1.4 Importance of ES in Terms of Nature and Landscape Protection in Slovakia

As mentioned above, sufficient area, proper structure and quality, especially of woody vegetation, is necessary for sufficient provisioning of ES air quality regulation. At the same time, this ES also contributes to the value of protected areas in terms of providing basic conditions for the life of organisms, including humans. Polluted air causes annual damage to health and premature deaths in many cases. A

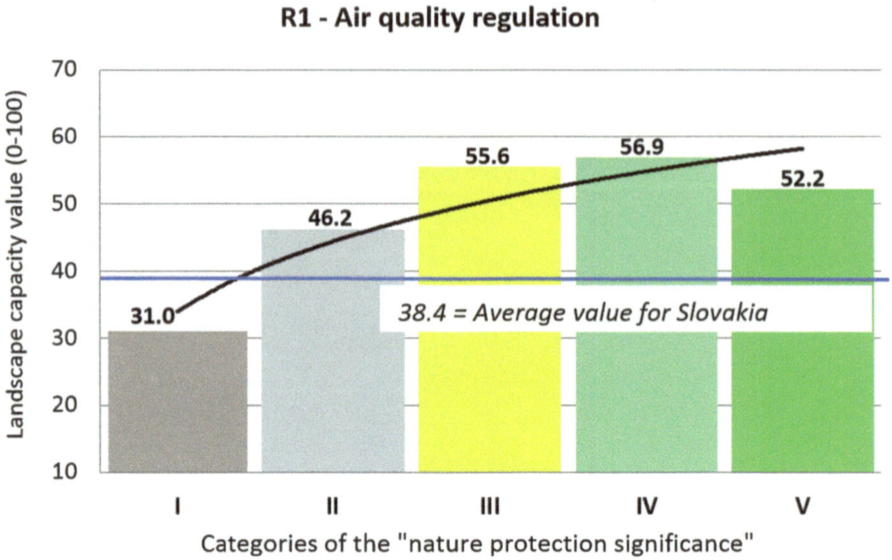

Fig. 4.2 The relationship between ecosystem service R1 and the significance of the territory of Slovakia in terms of nature and landscape protection

quality ecosystem included in protected areas contributes to the potential of a given area in terms of improving the conditions for life, and from the local point of view, it provides visitors of such a protected area with health benefits. The beneficial health effects of clean-air forest areas have been known for a long time and have been used for a long time in the form of treatments, thus increasing the credit and justification of individual protected areas with a high proportion of forest ecosystems. It should be noted that the most of protected areas in Slovakia has a majority of forest ecosystems, and it is therefore evident that in addition to the basic functions related to habitat and species protection, these areas also fulfil the function of ensuring/improving human health through production, regulation and purification of air for the Slovak population. The positive relationship between nature protection and the provision of the R1 regulatory service is also evident from Fig. 4.2, especially in categories III–V, where the capacity of the territory is highly above average.

Support for good-quality provision of ES air quality regulation is based mainly on the appropriate management of existing forest and woody areas in the landscape and in the planting of new green areas – especially in cities where demand for this service is the largest. Such measures are in most cases also supportive in terms of nature and landscape protection. The importance and function of this regulatory service in built-up areas needs to be emphasized – preserving and developing urban *green areas* will contribute to increasing air quality and the quality of life of inhabitants. Trees and other plants are involved in the removal of pollutants, which accumulate on the earth's surface due to dry deposition (Burkhard and Maes 2017). Urban green areas have many functions, but air quality improvement is one of the key ones. Therefore, green park areas and other areas with residential vegetation should be effectively protected.

4.1.5 ES Assessment for the Territory of Slovakia

Air purification and related microclimatic function is considered one of the important non-production functions and services of forest ecosystems. Čaboun et al. (2010) insert this function among the so-called atmospheric functions of the forest and consider the appropriate land use, good-quality forest structure and location of the area in terms of demand for this function, in particular, to be important factors of efficiency.

Various international assessments show that the most important provider of air quality regulation services is the forest ecosystems, which also applies for Slovakia. Especially important are forests with the natural species composition of trees (Fig. 4.3). In our conditions, it is possible to define the highest quality groups of forest habitats, which provide the ES air quality regulation – these include mainly oaks, hornbeams and scree forests. In terms of quantity, these include beech and fire-beech forests. In a smaller but qualitatively significant extent, the areas of

Fig. 4.3 A typical commercial forest dominated by European beech (*Fagus sylvatica*) is a good example of providing this ES in Slovakia. (Author: J. Černecký)

Table 4.1 Input data for capacity, demand and flow assessment of ES air quality regulation

Input data/ES	R1: air quality regulation
Capacity	Map of current landscape structure – reclassification as appropriate for ES provision
	Species composition, structure and condition of forest areas and stands (classification, types of stands, age of forests)
	Biomass volume in the landscape – leaf area index (LAI 2018)
Demand	Air quality in the region – polluted areas, concentrations of main pollutants
	Population of the municipality/region
	Recreation areas, special demand areas
Flow	Real effect of vegetation – rate of improvement of air quality
	Number of residents within the effect of ES provision

non-forest woody vegetation in the landscape (small forests, groves, shrubs, riparian vegetation), orchards and city parks also contribute to the provision of this service.

In order to assess the capacity and real provision of this ES, it is necessary to use data on the natural and real state of forest areas and the use of non-forest areas (Table 4.1). The pilot assessment at the national level was conducted using the appropriate and available data – especially data on forest areas (ESFT, stand types, age) and the current landscape structure of Slovakia. As a supplementary indicator for the volume of biomass, the so-called leaf area index (LAI) was taken into account from the European RS Copernicus system database (available online: www.copernicus.eu/en). These data were subsequently reclassified in a similar way as in the case of other ES to the relative landscape capacity scale for the provision of ES air quality regulation. The assessment result is shown in Fig. 4.4.

Table 4.1 also shows the basic indicators which can be used for future expression of the level of demand for ES as well as its real use. Logically, the highest demand for this ES is in the built-up areas, where the production of this service is the lowest. The number of inhabitants living in a particular territory is a suitable indicator. Demand can be expressed, for example, by the need to regulate air quality (delimitation of polluted areas) or the existence of special types of territories requiring improvement of air quality (zones, protected areas, etc.).

The ES flow is conditional on its real use – i.e. the level of air quality improvement by ecosystems, the number of inhabitants living in the affected area and the like. Obtaining such data at the national level is likely to be problematic, so it will be appropriate to use certain substitute indicators, so-called proxy indicators. Albert et al. (2016), for example, mentions the following as appropriate indicators of demand and consumption: the population density, the extent of settlement and exposure to air pollutants and to the harmful effects of urbanized environments.

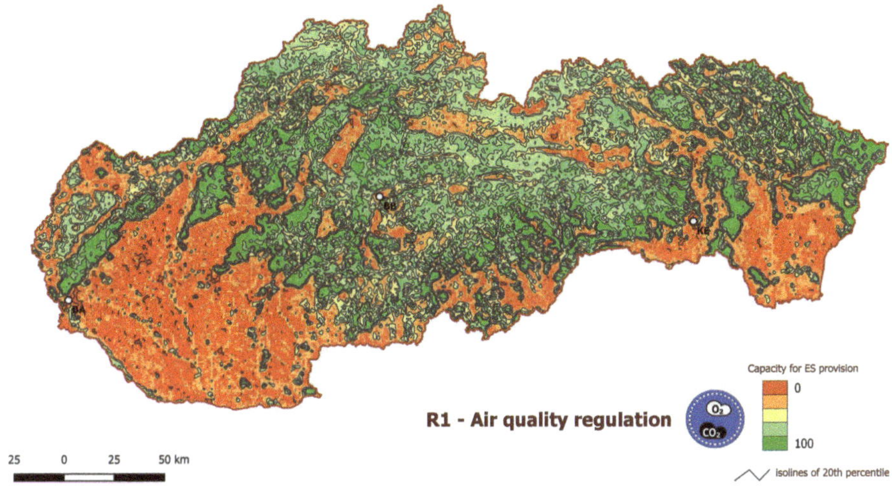

Fig. 4.4 Capacity of the landscape to provide ES air quality regulation

4.2 Water Quality Regulation (R2)

4.2.1 Definition and Brief Characteristics of ES

Water, as the basic prerequisite for life on Earth, provides a lot of different ES to people and at the same time supports the provision of all others ES (Coates et al. 2013). In addition to provisioning (drinking water and freshwater supply) and cultural services (recreation, healing), water represents a particularly important regulatory ecosystem service: for example, the correct timing and seasonal distribution of water supply and watercourses or water purification (in terms of water quality, including biological treatment as well as sediment storage, etc.) (Dudley and Stolton 2003; Bruijnzeel 2004; Brauman et al. 2007).

In order to define this ES and for its valuation, it is necessary to focus on what is understood by water quality. Water quality is often not exactly interpreted as the final ES. In this case, the connection between the provided ES and human well-being is particularly pronounced, as the water quality is highly perceived and highly valued by the public (Keeler et al. 2012).

In general, water quality may be seen as a set of several different biophysical parameters, which may affect the final provision of ES. These parameters can include, for example, the amount of nutrients (especially nitrogen and phosphorus), the acid-base balance, the presence/concentration of organic pollutants, pathogens, pesticides, industrial and pharmaceutical products, retained sediments, water colour, transparency or temperature (Smith et al. 2012).

Water purification for drinking and other purposes, as well as removing microbes and other toxins, is an important contribution to human health. The main components of ES water quality regulation identified based on the work by Smith et al. (2012) and UK NEA (2011) include, in particular, nitrate absorption, phosphorus absorption, regulation of pathogens and organic pollutants, sediment absorption and absorption of particulate organic carbon (POC), regulation of dissolved organic carbon (DOC), acidity balancing, water temperature balancing, pollutants dilution, prevention against the reproduction of harmful algae, the decomposition of organic pollutants, the intake of plant and microbial nutrients and the infiltration of pollutants into the soil and sediments. These processes contribute to the final ES, including the regulation of pollutants in other media, the provision of drinking water, fishing and recreation.

Given the number of subprocesses which make up this regulatory function, it is quite difficult to generalize the main factors of its functioning, as well as exactly capture the role of ecosystems. In general, however, ecosystems have the highest potential to regulate those water quality components which are bound to the sources of water collected in the river basin or the related water retention processes (Smith et al. 2012). They are therefore closely related to other regulatory ES, such as air quality and soil quality regulation, climate regulation and nutrient retention (UK NEA 2011).

4.2.2 Methods Used to Assess and Identify ES

The assessment of water quality regulation is quite challenging – the change in water quality affects several aspects of human well-being, and, moreover, benefits and/or costs can reach different groups of affected parties at different times and locations. Compared to other services, water quality regulation is thus much more complex. It cannot be assessed simply by one indicator or parameter, such as in the case of carbon sequestration (tons of captured CO_2). Similarly, the expression of marginal value may also be complicated, since any improvement in water quality by one degree can only affect the local level, and this value may vary depending on the spatial context and may have significantly decreasing marginal benefits (e.g. additional reduction of lake pollution by nutrients will bring only minimal additional benefits, and these benefits are also influenced by the state and proximity of other lakes). The time aspect can also play a very crucial role – current interventions can affect water quality for a relatively long period of time into the future, which complicates forecasting future values (Keeler et al. 2012).

Given the complexity of this ES, its assessment requires the use of an integrated approach and a combination of multiple assessment methods – in particular biophysical and economic methods. This approach makes it possible to capture the change in service provision in the case of, for example, changes in ecosystem management or land use, which can cause changes in water quality and thus influence the provision of ES and their value (Keeler et al. 2012).

Biophysical models link the changes in the landscape (ecosystems) with the change in water quality, as measured by, for example, the change in nutrient concentration, sediment deposition or input of chemical substances. Different models can be used for such assessment, such as SWAT (*Soil and Water Assessment Tool*) or InVEST. Outputs from these models can be expressed using the nutrients captured in the landscape or the loads at specific river basin endpoints. Similarly, biophysical assessment can be used to link water quality changes with the change in ES provision and goods, which directly affect human well-being.

In the case of aquatic ES, hydrological models and supporting indicators can also be used to capture complex interactions between different factors (climate, topography, geology, etc.). In their work, Grizzetti et al. (2016) divided these indicators on the basis of whether it is the ecosystem's potential/capacity to provide the given ES, the flow of the service or the social benefit.

Simpler methods of biophysical assessment focus on one or several key indicators. Pérez-Soba et al. (2015) mention the following as the most frequently used indicators: for example land use, hydrogeological properties, soil quality and vegetation properties – its spatial structure (canopy cover, biomass volume), naturalness, diversity and nutrient cycle. They are all included among the so-called proxy indicators which explain the operation and level of provision of the given ES only indirectly. Maes et al. (2014) and Czúcz et al. (2018) also stress the importance of qualitative indicators of water – organic carbon content, microbial activity, nutrient content and content of dissolved solids. The biophysical methods can also include the spreadsheet method/GIS-based approaches (Burkhard et al. 2012; Vihervaara et al. 2012).

Most of the national assessments of this ES also use biophysical indicators – the presence and quality of habitats significant for water purification (Denmark, Finland, Germany, Ireland, Luxembourg, United Kingdom) – or water quality indicators (Germany, Italy, Romania).

The aim of the economic assessment is then to reflect on how the change in the provision of ES will be reflected in its value and the benefits which people derive from it. To do this, different approaches can be used – a cost-based approach, whereby the estimation is focused either on the damage-cost avoided if the water quality is improved or on the costs associated with the increased health risk due to poor water quality. For the economic assessment, the so-called stated preference methods are used, where respondents directly answer the question of how much they would be willing to pay for some improvement in water quality. The third approach often includes the revealed preference methods, which, for example, compares respondents' willingness to pay for real estate near a good-quality water resource (Keeler et al. 2012; Grizzetti et al. 2015).

Various social assessment methods (usually combined with other assessment methods) can also be used to assess this regulatory ES, which also take into account social preferences attributed to, for example, drinking water (Perni et al. 2012).

4.2.3 The Main Types of Landscape and Ecosystems Which Provide ES

Water quality regulation is primarily linked to different types of aquatic ecosystems – i.e. lakes, rivers, marine and coastal waters, groundwater, freshwater and coastal wetlands, coastal areas and floodplains (Grizzetti et al. 2015). However, terrestrial ecosystems also play an important role, for example, in regulating the transfer of dispersed contaminants into the surface waters, particularly by infiltration and retention of pollutants in the soil (Smith et al. 2012). At the same time, in the case of watercourses, ecosystems in the upper parts of the basin have a major impact on water quality regulation (Fig. 4.5) – they dilute pollutants from point sources of pollution entering the aquatic ecosystems in lower parts of the basin in order to mitigate the impact of pollution on water resources (Smith et al. 2012). The main media for

Fig. 4.5 Important mountain watercourses usually have high water-purification capacity (TANAP, Javorová dolina). (Author: D. Kaisová)

the proper functioning of this regulatory ES include vegetation, soil and biota and wetland ecosystems (metabolic activity of plants and microorganisms).

It is therefore evident that not only the aquatic and wetland ecosystems themselves are *carriers* of ES water quality regulation, but the overall quality of local and regional ecosystems is also important, in particular sufficient extent, the appropriate spatial structure and quality of key types of ecosystems. In particular, we can include the forest ecosystems, wetlands and riparian vegetation, as well as permanent grassland near the waters and in river valleys.

The importance of this regulatory ES is also evident in the higher altitude and windward areas, where major water quality problems are often related to the deposition of atmospheric pollutants (sulphur, nitrogen, metals) as well as the colour/transparency of water in relation to dissolved organic carbon (Smith et al. 2011).

4.2.4 Importance of ES in Terms of Nature and Landscape Protection in Slovakia

Regulatory ES, including water quality regulation, are essential for nature and landscape protection. They create the conditions necessary to provide provisioning ES which bring direct benefits to people, such as crop production, availability of clean water and others, as well as cultural ES. However, compared to provisioning ecosystem services, changes in the provision of regulatory ES are reflected in a much longer time frame. There is, therefore, a risk that the deterioration of regulatory ES will not be seen immediately after the intensification of use of provisioning ES (Kumar 2010). Thus, the reduction in the ability of ecosystems to regulate water quality may be delayed. A frequent trade-off (i.e. provision of one type of ES at the expense of another service) is the more intensive use of provisioning ES at the expense of regulatory and cultural services (Rodríguez et al. 2006; Raudsepp-Hearne et al. 2010; Maes et al. 2012).

In the context of biodiversity protection, water quality regulation acts as a synergy – good water quality promotes water and water-related biodiversity at the same time (Smith et al. 2012). In a broader context, effective water quality regulation enables the healthy functioning of other ecosystems. On the other hand, in view of the close links between the different regulatory ES, deteriorated water quality may, for example, result in deterioration of soil quality and hence its ability to provide different soil-related ES (Smith et al. 2012). Water quality deterioration can have widespread ecological consequences – for example, water acidification, which has led to losses of biodiversity and fish stock over the past decades, which in turn has negatively affected the provision of recreational and provisioning ES (Smith et al. 2011).

According to the results of the assessment of this ES for the territory of Slovakia, the direct correlation between the capacity of the landscape and the significance of the territory in terms of nature and landscape protection is not as obvious as for some other regulatory ES. However, except for the highest degree of significance (V.), this correlation exists (Fig. 4.6) – particularly in degrees III–IV, there is the

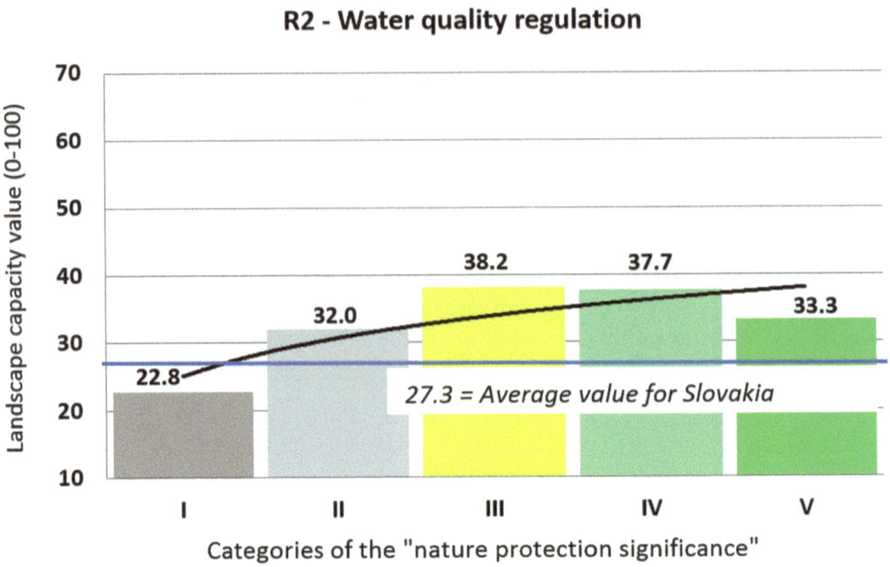

Fig. 4.6 The relationship between ecosystem service R2 and the significance of the territory of Slovakia in terms of nature and landscape protection

highest share of forest ecosystems, which are crucial for the provision of this ES, together with hydric ecosystems.

Maintaining aquatic ecosystems and especially wetlands is essential for the sustainable provision of water quality regulation (Fig. 4.7), both in terms of quantity and quality – not only inside but also outside of protected areas. This will necessitate focusing on the factors which are now most affecting the water quality, notably agriculture, industrial pollution and land use management (MEA 2005, Smith et al. 2012). In our conditions, we can also include residential development and the development of technical infrastructure (such as Žitný ostrov). Measures which can mitigate the impact of stress factors on water quality regulation include, for example, the development of buffer zones, which provide biological continuity between rivers and their riparian zones and, where possible, use green infrastructure, such as restoring coastal areas, wetlands and water retention areas that promote biodiversity and soil fertility and prevent flooding and droughts (European Commission 2012).

4.2.5 *ES Assessment for the Territory of Slovakia*

Hydric ES in Slovakia was studied, for example, by Bujnovský (2018), who estimated, among other things, the value of the regulatory ES on the example of the valuation of nitrogen retention in the aquatic environment – but only for the whole territory of Slovakia based on the value transfer from an analogous study (estimate of 3 million EUR per year). However, this is only a partial assessment, with the

Fig. 4.7 River Turiec with important aquatic habitats and flowering macrophytes (*Batrachion fluitans*). (Author: J. Černecký)

value being very low. Another specific assessment of this ES for Slovakia is not known. There are only partial studies assessing some aspects of water quality regulation, for example the ability of the soils to immobilize and transform risk chemical elements (Vilček 2014) or theoretical elaboration of water-protection function of forests (Čaboun et al. 2010; Konôpka 2012).

Považan et al. (2014a) propose to use the ecosystem infiltration capacity (e.g. water volume/surface area) as indicators of ES water regulation – volume per unit area/per time; water retention capacity (in mm/m^2) or water retention capacity by alluvial meadow (in mm/m). In the case of water purification, they present the nutrient capture by wetlands (tons or percent): water quality in aquatic ecosystems (sediments, turbidity, phosphorus, etc.). This represents the use of biophysical indicators, which are, however, more appropriate for the local to the regional level.

For a simple assessment, it is possible to use some of the methods which are used in foreign studies, for example models like InVEST or ESTIMAP, or the basic screening method of the so-called Burkhard matrix. Consequently, it is possible to follow up the biophysical assessment with an economic assessment.

For the initial assessment of the capacity of the landscape of Slovakia to provide ES water quality regulation, available data on the current land use, quantity and quality of vegetation focusing on forest areas as well as data on soils (absorption capacity) and relief (slope gradient) were used – see Table 4.2. The basic indicator used was the regulatory capacity of vegetation (the result of the R1 regulatory

Table 4.2 Input data for capacity, demand and flow assessment of ES water quality regulation

Input data/ ES	R2: water quality regulation
Capacity	The regulatory function of vegetation – result of ES R1 assessment
	Slope inclination – coefficient of runoff attenuation
	Soil absorption capacity
Demand	Water quality in the territory – contaminated areas, concentrations of main pollutants
	The population of the municipality/region in the demand areas
	Special areas of demand (water management, fishing, bathing, recreation)
Flow	The real effect of ecosystems – the rate of water quality improvement
	Number of residents within the effect of ES provision, the attendance of the affected areas

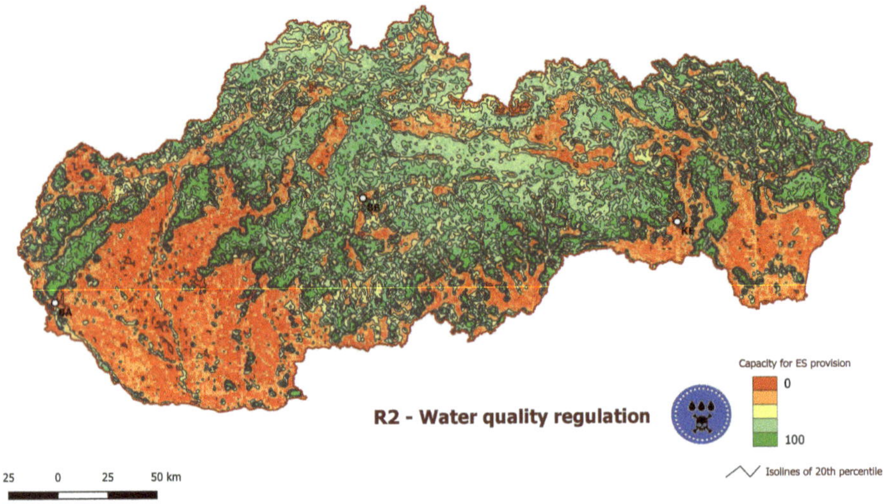

Fig. 4.8 Capacity of the landscape to provide ES water quality regulation

service assessment); the soil and relief correction coefficient was used to refine the input value. The values obtained were expressed in relative scale – the result of the assessment is shown in Fig. 4.8.

Together with the indicators for expressing the landscape's capacity, Table 4.2 also lists the indicators suitable for determining the level of ES demand and its real use. A suitable demand indicator is, for example, the need for water quality regulation – defining the polluted areas. The number of inhabitants living in a particular area (whether in polluted areas, but also, for example, in districts or in municipalities) is also important. Areas of demand also include areas of activities requiring clean water (recreational areas, watercourses and reservoirs, fishing grounds, etc.).

The real use of ES water quality regulation is determined by the quality of the water itself (watercourses, reservoirs and groundwater areas with good quality), an improvement over a certain period of time, the number of inhabitants living in such a territory, attendance and the like. Population density, spatial distribution of settlements, recreation or changes in the quality class of watercourses may be used as proxy proxy indicators similarly to ES R1.

4.3 Erosion and Natural Hazard Regulation (R3)

4.3.1 Definition and Brief Characteristics of ES

Erosion is a relief-forming process caused by the effects of exogenous processes, which lead to the removal of topsoil cover (usually topsoil), often faster than soil formation. The erosion processes occur mainly due to water and wind and can lead to a qualitative deterioration and loss of productivity, especially in agricultural areas, which can have serious consequences for the costs of agriculture and food. Analysis of erosion-sensitive areas allows decision-makers to anticipate this risk and implement erosion reduction measures which can be achieved through preventive land use and management (Becerra-Jurado et al. 2016).

The ES of protection of the territory against these adverse processes is based on the ability of the river basins to determine the surface water runoff in the landscape so as not to damage natural resources. Relief and land use method establish the basic framework for the regulation of processes associated with surface runoff and water retention. The condition of soil saturation leads to processes of surface erosion, gully erosion and slope gravity disturbances. Soil erosion is a natural and normal process – it is important to ensure that the soil runoff limits are not exceeded during land use (e.g. in the Universal Soil Loss Equation (USLE) model). This ES is indirectly related to other ecosystem land services, such as carbon sequestration, biomass production, water quality control, nutrient and contaminants filtering and their retention (Palm et al. 2014; Vilček 2014).

Several types of soil erosion are recognized – the basic types include surface water erosion, gully erosion and wind erosion. The specificity of the last one is that it is not so heavily connected to rainfall-runoff conditions and relief. It depends more on the soil characteristics (texture, structure, moisture regime), land-use methods and wind conditions.

The second group of processes associated with rainfall-runoff conditions in mountain and foothill areas includes *slope deformations – landslides* (geodynamic phenomena in the broader sense). It is a relatively fast gravitational transfer of slope masses (top layer of soil, debris and rocks along the so-called shear surface) from the source area along the slope, which results in the deformation of the original relief and the creation of a new form of relief (the landslide itself) usually composed of erosive, transport and accumulation part. Slope deformations have different causes and different characteristics, and therefore there are different types of landslides (available online: www.usgs.gov). Ensuring slope stability is quite complicated, which is related to the properties of the geological bedrock and slope inclination as passive factors and to the active action of rainfall, possibly to land use and technical human intervention in the country.

Relief-forming processes in the conditions of the alpine landscape are also associated with risks due to the movement of snow masses in the form of avalanches. The main *avalanche forming* factors include the height of the snow cover, the air temperature regime and the morphometric attributes of the relief. In addition to moving snow, avalanches have the ability to disrupt the vegetation and soil-substrate cover. Most often, the source parts of avalanches are at risk in this respect, so we use the potential avalanche formation model to identify and assess these. Avalanche disturbances also have a serious impact on the forest ecosystems of the subalpine and montane levels.

ES regulation of slope processes associated with the movement of material is clearly related to the rainfall-runoff regime in river basins and is limited by land use and the protective effect of vegetation, including types of ecosystems able to retain water in the landscape. The spatial structure of vegetation and its properties play an important role in soil protection and slope stabilization – for example plant roots help to stabilize the soil, minimizing soil degradation and also decreasing the sediment in watercourses and thus contributing to better water quality (Preston and Raudsepp-Hearne 2017).

To conclude, ES erosion and natural hazard regulation is understood as the *ability of ecosystems and landscape to regulate adverse relief processes* – especially to prevent and mitigate water and wind erosion, landslides and selected gravity processes and, to some extent, avalanche risk.

4.3.2 Methods Used to Assess and Identify ES

Relief processes associated with soil erosion, slope processes and avalanches are assessed on a long-term and global basis with the prevalence in the use of biophysical models based on natural environment parameters and landscape use factors.

The most commonly used indicators of soil erosion assessment include, in particular, the landscape use, relief (inclination), mapping of real processes (landslides and erosion occurrence), soil parameters (depth, texture, retention capacity) and vegetation characteristics – especially location, cover and spatial structure (Pérez-Soba et al. 2015; Czúcz et al. 2018).

In the national assessments of this ES, only the water erosion processes are virtually investigated within Europe. The properties of vegetation are almost always used in the calculations – the area of individual elements (especially forests and protective stands) and their spatial representation, to a lesser extent the properties of soils (Finland, Germany, Romania) and relief (Romania, UK). In Italy, the InVest model was also used to assess the territory's erosion protection (Giarratano et al. 2018). In the available national ES assessments, wind erosion, slope processes and avalanches were not considered at the national level.

Generally, the most widely used assessment method for ES erosion and natural hazard regulation is modelling – a wide range of models is used, mainly to calculate the *potential and actual water erosion* (Markov and Nedkov 2016). These include various modifications of the USLE, RUSLE (Revised Universal Soil Loss Equation) and USPED (Unit Stream Power-based Erosion Deposition) models in the GIS environment (for an overview see for example Šinka et al. 2013). The SSCRI modelling tool was developed for Slovakia's agricultural territory (Antal, 2005 – available online: www.podnemapy.sk/erozia/).

The calculation of potential and actual *wind erosion* is also mostly based on modelling, using models in the GIS environment (e.g. WEQ, Wind Erosion Equation; TEAM, Texas Tech Erosion Analysis Model; AUSLEM, Australian Land Erodibility Model – review by Grešová 2010). For Slovakia, it is possible to use the classification of the risk to the territory of Slovakia by wind erosion (overview in, e.g., Streďanský et al. 2005; Kobza et al. 2005, application within the portal available online: www.podnemapy.sk).

Minár and Tremboš (1994) present the method of determining *gully erosion* as a manifestation of concentrated surface runoff – it is based on the attributes of the relief slope, the slope length and the rock resistance factor. The authors also developed an empirical formula to determine the threat of gravitational *slope deformation* activation. A model for assessing the susceptibility of the area to landslides DYLAM (Pechoušková 2006). An antegrated assessment of natural threats was investigated by Šabo et al. 2012 and others.

The basic model for spatial identification of *formation of avalanches* used in Slovakia is the model of avalanche threats (Hreško 1998; Barka and Rybár 2003; Žiak 2012 and others).

4.3.3 The Main Types of Landscape and Ecosystems Which Provide ES

The most important type of country that provides the ES associated with the regulation of the effects of slope processes in Slovakia is represented by *forested parts of hills, highlands and mountainous areas*, while their real effect is determined by the local characteristics of the relief, climate and hydrological conditions (rainfall-runoff conditions). It is the vegetation cover that is the determining factor that can prevent most of these processes from occurring – the anti-erosion effect of vegetation is the most important (Fig. 4.9). The greatest effect comes from vegetation in the case of a suitable spatial structure (wood cover) and quality (species and age

Fig. 4.9 Forest ecosystems protect the steep slope against landslide and erosion and provide a rainfall retention function, thereby protecting property and health (Horná skala, Malachov). (Author: J. Černecký)

varied vegetation, with developed undergrowth), where it can be a direct limiting factor for such processes. When comparing the different types of forest stands, deciduous and mixed forests are clearly more favourable, while, for example, spruces are more susceptible to the appearance of processes due to their worse structure, lower stability and stand resistance.

On the other hand, in case of sudden events caused by extreme factors (extreme precipitation, slope stability erosion, earthquake), not even a very good and quality vegetation structure can guarantee the prevention of the occurrence of such an event – for example, catastrophic landslides or erosion phenomena occur periodically in forest areas. However, the most common cause of such events is the *unfavourable impact* of human activities in the area (construction, transport, mining of raw materials, deforestation, etc.).

A very good anti-erosion effect is also provided by permanent grassland – meadows of various types. That is why the resistant type of landscape is formed by the varied structures of the submontane agricultural landscape with a prevalence of grasslands and a high proportion of permanent vegetation, especially in the case of preservation of historical structures of the agricultural landscape (especially terraced and narrow-banded fields and meadows with a limit – Špulerová et al. 2017).

The most risk-prone types of landscape in terms of susceptibility and occurrence of erosion processes include the intensively used agricultural land with the dominance of large-scale arable land. It is in this type of territory that the most common manifestations of water and wind erosion occur. In these territories, the main ES caused by the occurrence of slope processes is, in particular, the protection and gradual restoration of soil productivity, in particular by respecting anti-erosion measures, appropriate forestry and agricultural practices and good agricultural practice and integrated nutrient provision management.

4.3.4 Importance of ES in Terms of Nature and Landscape Protection in Slovakia

According to the assessment of the territory of the SR from the point of view of ES erosion and natural hazard regulation and other slope processes, the relationship between this ES and the significance of the area is clear in terms of nature and landscape protection, with a significant positive correlation (see Fig. 4.10). In the previous text, the importance of forests and extensively used agroforestry areas, which form the foundation and majority of the area of the protected nature areas in Slovakia, is emphasized.

In addition, the protection and management of protected areas create a prerequisite for regulation of slope processes by eliminating, limiting or conditioning human activities at various stages of protection which could trigger or accelerate the considered morphodynamic phenomena. The parts of the protected areas themselves or their ecosystems thus contribute to the elimination of the emergence and development of processes which could change the functioning conditions of the ecosystems concerned.

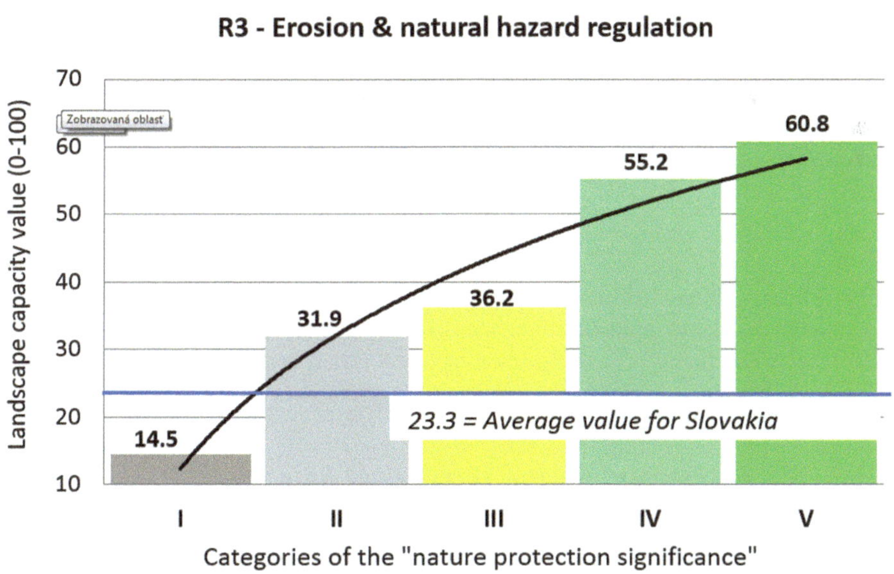

Fig. 4.10 The relationship between ecosystem service R3 and the significance of the territory of Slovakia in terms of nature and landscape protection

Fig. 4.11 Limits and ecotones in the traditional management of Hriňovské lazy create important elements for preventing soil erosion and landslide after torrential rains. (Author: F. Petrovič)

Such a *synergistic* effect also applies vice versa – the management of commercial activities of people focused on the prevention, elimination or mitigation of the effects of relief-forming processes is also a supporting factor for more effective nature protection. The protective measures include, in particular, finer and natural farming and agricultural methods (Fig. 4.11), which are now also strongly supported by the EU subsidy system (including, e.g., agri-environmental measures and agroforestry systems). Nevertheless, these measures in the current setting are not yet sufficient, and their implementation in practice lags behind the theoretical basis, while it is necessary to significantly change/set the schemes so that they can really contribute to the protection of nature in Slovakia and not vice versa.

On the other hand, some geomorphological processes can be closely linked to the development of important ecosystems and can be a forming active factor in their functioning. Avalanche ecosystems in the alpine environment of high mountains (Hreško and Bugár 1999; Fischer et al. 2012), wetland ecosystems in non-draining depressions of landslides, habitats in watercourse channels after flood events and so on are perceived as such. Many original disturbances like that are part of protected areas, in some cases, they were directly one of the main factors for their declaration.

4.3.5 ES Assessment for the Territory of Slovakia

Unlike most other ES, the issue of erosion and other slope movements is very well investigated and identified within the territory of Slovakia. When it comes to *water and wind erosion*, it is assessed regularly for agricultural land in the form of maps of potential and current water and wind erosion within the SSCRI Information

Service (available online: www.podnemapy.sk/default.aspx). Among other things, the portal provides the possibility of interactive modelling of current water erosion at the local level. According to the SEA (Enviroportál 2018) assessment in 2017, there was 38.6% of agricultural land (761.6 thous. ha) at risk from water erosion. While for the hilly landscape middle category of potential erosion (4–10 t/ha/year) is typical, for more rugged sub-mountainous and mountainous areas it is particularly the high threat (10–30 t/ha/year) and to a lesser extent extreme threat (above 30 t/ha/year). The area of soils potentially affected by wind erosion in Slovakia in 2017 amounted to 6.7% (131.6 thous. ha) of agricultural land. This type of erosion is associated with the lowland areas of Western, Southern and Eastern Slovakia. Water erosion on forest land is not assessed in this way, although several authors have processed maps of potential and actual water erosion of Slovakia according to the above-mentioned models, based on variously detailed data (Antal 2005; Gallay 2010 and others).

Slope deformations are inventoried and assessed for the territory of Slovakia within the competence of the State Geological Institute of Dionýz Štúr (SGIDŠ) in Bratislava, which operates a database of slope deformations (available online: www.apl.geology.sk/geofond/zosuvy/) and Atlas of Slope Stability Map of the SR (available online: www.geology.sk/geoinfoportal/). There are 21,190 slope deformations registered in Slovakia, and these occupy an area of 257.5 thous. ha, which represents 5.25% of the territory of Slovakia. The largest number comes from landslides with registered number of 19,104, accounting for 90.2% of all registered slope deformations. The expansion of slope deformations is associated mainly to areas built by Paleogene rocks and Mesozoic klippen belt and Paleogene of the outer flysch belt. Approximately, 12% of the total number of landslides in Slovakia is active.

Slope deformations represent a phenomenon which significantly affects the state and effective use of land. It acts as a constant threat where buildings are located without adequate measures and repeatedly causes damage to the land, line and other structures, underground and overground utility networks, as well as agricultural and forest land. The landslide risk in some regions of Slovakia is also currently increasing due to the intensified direction of construction activity from flat and slightly inclined areas to sloping and more exposed areas. This trend is particularly evident in the villages of mountainous regions of Slovakia. It is caused not only by the lack of suitable building plots in flatlands but often also by the targeted placement of buildings on slopes due to the attractiveness of the environment. The classification of *avalanche risk* in the mountains of SR is realized through the portal of the Mountain Rescue Service (available online: www.laviny.sk) and the GIS portal of the alpine environment of the SR (available online: www.avalanche.sk). Similar to other alpine areas, avalanche processes are associated with the alpine, subalpine and supra-montane zones in Slovakia. Their disturbing modes according to Bebi et al. (2009) perform bidirectional interactions in which avalanches affect the structure and composition of the forest and avalanches affect the structure and composition of the forest. The occurrence of avalanches is associated only to the high mountains of the Carpathians, predominantly above the top boundary of the forest. The creation of detailed avalanche maps of Slovak mountains is currently under work by, for example, Žiak (2012).

The occurrence of individual types of geomorphological processes and susceptibility to them is elaborated relatively in detail in Slovakia. However, a comprehensive assessment of slope processes and, in particular, the degree of protection against their effects (which represents the ecosystem service itself) is not yet present – although it was methodically investigated and applied in the model area by, for example, Šabo et al. 2012.

For the purposes of the ES catalogue of Slovakia, the calculation of ES water erosion regulation was performed as a pilot assessment based on several available and created documents (see Table 4.3 for a list of background maps). The result is not presented in the *classical* form of the intensity of potential or actual water erosion – the capacity of the landscape is expressed as the protective effect of vegetation and ecosystems against the processes of erosion and other geodynamic phenomena (such as the difference between potential and actual erosion intensity). All the important factors of the erosion susceptibility of the area – the relief (the inclinations of the relief, the shape and the length of the slopes), the precipitation intensity, the soil characteristics and the nature of the use of vegetation and ecosystems – were included in the calculation.

It is logical that the highest protective effect of vegetation is typical for rugged sub-mountainous and mountainous areas with a high-water erosion susceptibility (Fig. 4.12) – mostly wooded and grassed areas. The moderate effect of vegetation is typical in lowland hills and lower mountain ranges, and low effect in flatlands (due to the fact that potential erosion is low in these areas).

Table 4.3 Input data for capacity, demand and flow assessment of ES erosion and natural hazard regulation

Input data/ ES	R3: ES erosion and natural hazard regulation
Capacity	Land use – CLS types
	Nature of vegetation – structure and quality of forest habitats (alternative C-factor)
	Relief – inclination, segmentation and length of slopes (alternative LS factor)
	Erosion susceptibility of soil (alternative K factor)
	Precipitation intensity (alternative R factor)
Demand	Potential water erosion and susceptibility to other processes (wind erosion, landslides, avalanches)
	Integrated assessment of the territory to adverse geomorphological processes
	Number of inhabitants of municipalities/areas in areas prone to assessed processes
	Definition of particularly sensitive areas – urbanized areas, recreational areas
Flow	Real effect of ecosystems – the measure of the protective effect of vegetation for individual processes
	Integrated flow assessment – ES utilization rate for all assessed processes
	The number of inhabitants within the ES reach – population protection

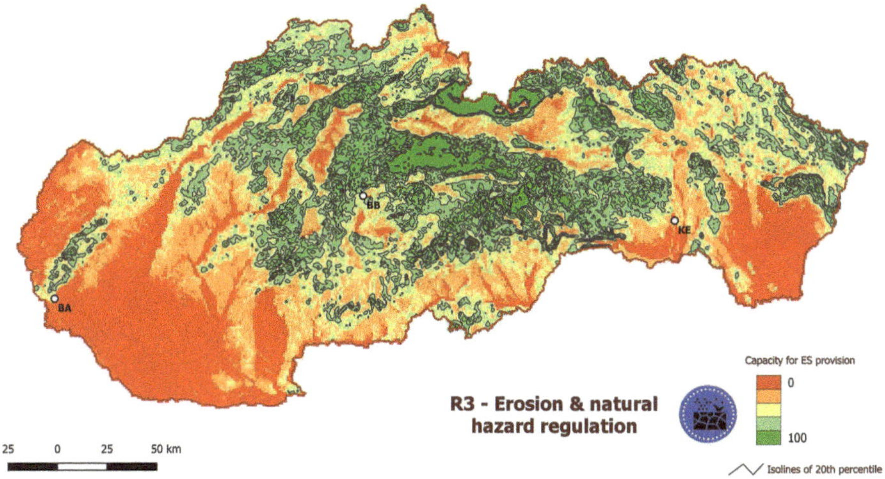

Fig. 4.12 Capacity of the landscape to provide ES erosion and natural hazard regulation

Table 4.3 also shows the indicators proposed or appropriate for determining the level of demand for a given ES and its real use. For the ES demand, it is appropriate to express the need for regulation of slope processes, based mainly on the definition of vulnerable territories and the determination of the population living in such territories. The territory of demand can also include sensitive areas (urbanized areas, recreational areas) characterized by the number of affected inhabitants.

The real use of ES erosion and natural hazard regulation can be assessed by expressing the real effect of vegetation and ecosystems in vulnerable territories as well as some integration of action in relation to other processes (not only erosion but also landslides and avalanches). It is also possible to express the number of inhabitants living in the territory with a real positive effect of the ES.

In the absence of the necessary data, proxy indicators can be used (e.g. population density, spatial projection of settlements and other activities).

4.4 Water Flow Regulation (R4)

4.4.1 Definition and Brief Characteristics of ES

Floods are complex events which are difficult to predict, with many factors contributing to their occurrence. Water retention capacity in river basins is therefore particularly important for flood risk reduction. Another important prerequisite for landscape protection is the infiltration capacity of the soil and the presence of habitats with high-water retention capacity. These habitats should be given particular attention. As floods can have a devastating effect in the landscape, the monetary and social benefits of adequate water flow regulation are enormous (Becerra-Jurado et al. 2016).

The assessed ES water flow regulation expresses the *river basin's ability to regulate water runoff during extreme rainfall events* so as to avoid flooding in the context of exceeding N-years flow rates and minimizing the duration of a flood event. With extreme flows in watercourse channels, there also exists a threat of waterlogging and flooding due to high groundwater levels.

Supporting natural water flow regime in river basins through natural ecosystems provides people with many benefits – for example by mitigating droughts and extreme flood events, mitigating extreme minimum and maximum watercourse flows and providing natural water supplies for utility purposes. Changes in landscape cover and land use can affect the timing and extent of flow, flood discharges and saturation of watered alluvial layers. Flood mitigation factors or water regime adjustments also include soil permeability, the presence of alluvia and wetlands, which may also reduce the need to build technical infrastructure (Preston and Raudsepp-Hearne 2017).

Based on the Millennium Ecosystem Assessment (MEA 2005), ES water flow regulation can be defined as the impact of ecosystems on the timing and extent of water runoff, floods and refilling of groundwater collectors, primarily in terms of ecosystem or landscape potential to collect and retain water.

ES water flow regulation can be comprehensively understood as an *ecosystem and landscape ability to regulate outflow processes* – especially to mitigate extreme volumes of surface runoff and flood discharges. It is suitable to assess different spatial levels – from defined micro-basins and reference profiles on watercourses through larger river basins to national levels. It is also appropriate to assess the real significance of the ES with the emphasis on the distribution of inhabitants with regard to the areas prone to the occurrence of flood events.

4.4.2 Methods Used to Assess and Identify ES

Similar to the case of erosion and other slope processes, biophysical methods and indicators are used in the assessment of runoff conditions and flood risk. The simpler methods include the use of various indicators and mapping methods; the more complex approaches include the use of various complex computational models.

A summary of indicators applied in various world studies is provided by, for example, Pérez-Soba, Harrison et al. (2015) and Czúcz et al. (2018). They highlight the indicators of land use (spatial structure of use, share of greenery), relief (inclination, size and shape of the basins), hydrological parameters (runoff, flow rates, occurrence of floods), soil parameters (retention capacity, permeability) and ecosystem properties (spatial structure of vegetation – coverage, distribution). Maes et al. (2014) also emphasize the importance of river floodplains and their threats and the proportion of water elements and wetlands in vulnerable territories.

The assessment of this ES has been largely carried out in the framework of national ES assessments in European countries. The most commonly used indicators include the area of water retention ecosystems and runoff mitigation (Finland, Germany, Ireland, Luxembourg, Romania). Some countries have also used models for water retention and water runoff from the river basins (Germany, Romania, United Kingdom).

More complex computational models for the calculation of surface runoff volumes in micro-basins and the interpretation of landscape properties with respect to their regulation are represented by, for example, HEC-HMS (Hydrological Modelling System) and HEC-RAS (River Analysis System) models. HEC-HMS serves to simulate the rainfall-runoff process, to calculate the volume of direct runoff volume from the area and to simulate peak flows on the basis of N-year precipitation (model description, e.g. Kadlec 2010; Jeníček 2009). HEC-RAS is a one-dimensional hydraulic model designed for flow modelling in river systems (e.g. capacity calculations of selected watercourse profiles – e.g. Černý (2012) and website of Hydrologic Engineering Centre USACE – available online: www.hec.usace.army.mil/).

The assessment of runoff conditions and flood risk is also investigated by several Slovak authors. Solín (2011) created a methodology not only to determine the hydrological balance but also to identify runoff genesis in the context of land use changes. An important contribution to ES assessment is the generation of integrated flood risk assessment models in basins (Solín et al. 2016; Solín 2017). For the processing of flood maps and watercourse risks, water depths and water flow rates for floods with a repetition time of 5, 10, 50, 100 and 1000 years have been specified.

An alternative calculation of the potential and real direct surface runoff from the basins and in the reference profiles is provided by the method of runoff curves (so-called CN curves), which has been prepared down to the level of micro-basins with the use of ArcGIS superstructures or other GIS systems – more details can be found in works by, for example, Smelík (2016), Šinka et al. (2013), Kaletová and Šinka (2012) and Gallay (2010).

4.4.3 The Main Types of Landscape and Ecosystems Which Provide ES

The ES water flow regulation needs to be assessed in the spatial context of hydrological systems of rivers and streams, which form the backbone of almost all types of socio-economic activities from urbanization, communication networks, agriculture and so on. Hydrological regulatory functions are similar to those of the previous ES, directed at the mitigation of erosion processes – the essence of which is the ability of a landscape and ecosystems to retain surface runoff, reduce its volume (which is also done by water consumption by ecosystems), slow down runoff processes and transform them as much as possible into subsurface levels of soil and subsoil.

The specific feature is that runoff processes and floods are manifested through the hydrological network of watercourses and their offshore systems; therefore, the status of aquatic ecosystems is crucial for this ES. The natural and well-functioning watercourses, their valleys and wetlands carry the key regulatory function – this includes the dynamic ecosystems which can best transform flood waves and high-water levels into lower parts of the basin. Unfortunately, such watercourses are limited in Slovakia to virtually only mountain basins, as they almost completely retreated from the structural basins and lowlands due to anthropogenic adaptations and commercial land use. This is particularly true in the case of wetland ecosystems, which were also of great importance in the lowlands, not only in terms of regulatory function but also in balancing the landscape's moisture deficit during the growing season (Fig. 4.13).

Forest ecosystems are the most important area element in the landscape with a water flow regulation function. Forests and permanent vegetation in the landscape (groves, line stands of woods) are key elements for rainfall transformation and runoff regime. Similar to the anti-erosion function, the appropriate spatial structure (especially the total biomass volume) and the quality of the stands are important. It is true that species and age-diverse stands with developed scrub and herbaceous undergrowth are more stable and more suitable in terms of runoff transformation and balancing. The problem is that disturbing the stability (especially of non-native) of forest stands in the mountainous areas of Slovakia, the frequent occurrence of calamities and the subsequent large-scale harvesting over the last 10–15 years reached an almost *catastrophic* extent, which largely undermines the fulfilment of forest regulatory functions.

In addition to watercourses, their shoreline vegetation and perennial permanent vegetation, permanent grassland is also important in the agricultural landscape. These permanent grasslands fulfil hydrological functions in addition to their anti-erosion functions. As with other regulatory functions, diversified land use patterns

Fig. 4.13 Flood areas in the ecosystem help significantly in flood protection (wetland near the village of Rad on the Východoslovenská nížina lowland). (Author: J. Hreško)

of mosaic nature, which mainly form the submontane agricultural land with preserved small-scale historical structures, have a high value in this respect.

On the contrary, ecosystems negatively affecting the fulfilment of regulatory hydrological ES include the intensively used agricultural and urbanized areas – with a predominance of settlements, technical elements and large-scale fields. Such a landscape is characterized by a changed hydrological regime not only of the agricultural landscape itself but also of the unsatisfactory state of watercourses and other hydric elements. Water management measures are also part of the land reclamation and amelioration measures – their main objective was to improve the state of the landscape in terms of increasing water availability and productivity (building of water reservoirs and other sources, irrigation, hydro-melioration channels, etc.) and protection against the undesirable effects of natural processes (watercourses modification, water flow). Unfortunately, in the second half of the twentieth century, a number of modification and interventions in the landscape were implemented, which had a considerable negative impact on the functioning of natural processes and mechanisms ensuring the fulfilment of hydrological regulatory functions. Technical buildings in the landscape require care and maintenance, which is often not the case for water structures. Therefore, instead of performing their original purpose, water management structures are severely limited in their function, and their construction has rather disrupted regulatory relations and processes, especially in the lowland and basin landscapes. While the large water management structures like dams and embankments can prevent floods in the lower basin areas, they even such structures are not able to prevent the occurrence of floods and flood damage in higher parts of the river basin. It is in these areas where floods are particularly frequent in case of poor landscape state and inadequate ways not only of urbanization but also of agriculture and forestry.

4.4.4 Importance of ES in Terms of Nature and Landscape Protection in Slovakia

As with most regulatory services, a landscape with well-functioning regulation of hydric processes is in line with the performance of natural protection functions. When assessing the relationship between the landscape's capacity to fulfil this ES and the significance of the territory of Slovakia in terms of nature and landscape protection, there is no direct correlation or positive correlation (see Fig. 4.14). This is probably due to the fact that the assessment is focused on sub-basins, not the ecosystem types themselves, and at the same time this ES is important and present not only in mountain areas (direct transformation and deceleration of runoff conditions) but also in lowland areas (flood prevention, water management etc.).

When assessing the landscape's hydric functions in relation to nature protection, it should be however emphasized that the method of landscape management and the implementation of possible water management adjustments are essential. It is

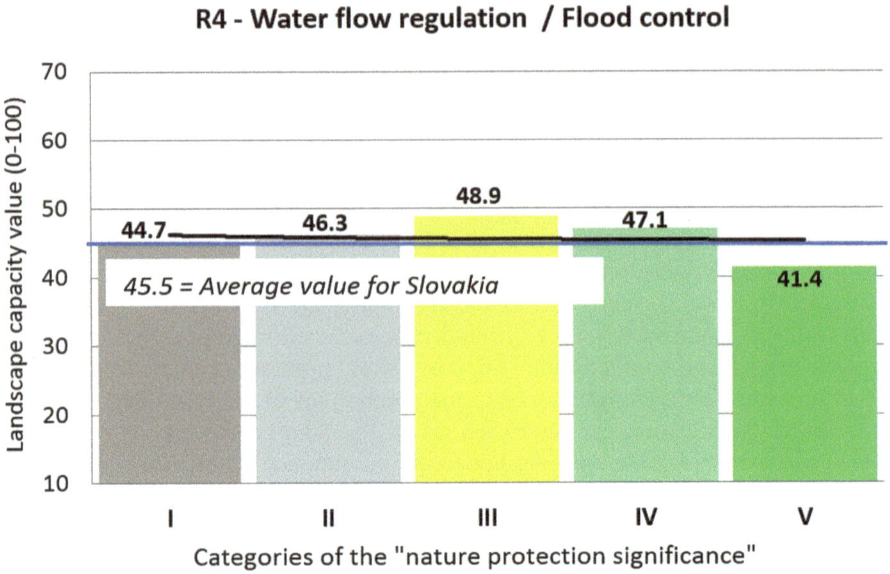

Fig. 4.14 The relationship between ecosystem service R4 and the significance of the territory of Slovakia in terms of nature and landscape protection

essentially a long-term dispute between the advocates of *nature-based* and *technical* solutions and measures.

Nature-based hydrological measures and river basin management are based on the preference of nation-wide measures to change the way the landscape is managed. It is a return to small-scale and diversified forms of agriculture, revitalization and renaturation of hydric ecosystems, planting of woody vegetation, local anti-erosion and flood control measures, natural forestry measures, small-scale and selective forest management and the like. Such measures are in line with most other landscape regulatory and support functions and services, including nature conservation. Reciprocally, effective protection and management of protected areas of nature is in line with such a concept of hydrological functions of the landscape.

Hydrotechnical solutions and measures are mainly represented by *hard* interventions in the landscape – building reservoirs, polders and flood-protection dams and regulating and straightening watercourses. Although these provide immediate solutions and can improve flood protection for large territories, they have negative consequences in terms of other ecosystem functions and landscape services – in many cases being very significant and irreversible, including nature and landscape protection. For that reason, such interventions are absolutely inappropriate in protected areas – and their implementation in other territories should be clearly justified by the inability to protect the territory by other means.

4.4.5 ES Assessment for the Territory of Slovakia

Extreme hydrological processes in river basins leading to periodic floods of local to regional extent are quite frequent in the territory of Slovakia. Their importance is increasing in the current climate context, so it is logical that relatively high attention is paid to mapping and assessing vulnerable territories both in the government area and in science and research.

The flood-threatened areas are very accurately expressed in the flood maps – the flood hazard map and the flood risk map of the watercourses of Slovakia. Maps are prepared, maintained and updated by the Slovak Water Management Company, š.p. as a tool to *reduce the adverse effects of floods on human health, the environment, cultural heritage and economic activity by reducing the extent of flooding, reducing vulnerability and mitigating the negative consequences of floods* (available online: www.mpompr.svp.sk). The maps show the territories threatened by various floods (Q_5 to Q_{1000}) and the data on the potentially adverse effects of possible floods on the population and the economy. They were created by simulating steady uneven water flow through a mathematical hydrodynamic model.

The long-term classification of the Slovak basins is performed also by the Geographical Institute of SAS. The basic classification of runoff regulation in river basins is provided by the results of the regional hydrogeography of Slovakia (Solín 2003, 2011). In addition, Solín et al. (2016) created five classes of the Slovak river basins according to flood risks, which can form the basis for assessing the need for ES regulation of runoff conditions.

Both of these and most of the other approaches assessing the territory of Slovakia are focused on mapping, or flood risk assessment, which in the ES context represents a demand-side and not a capacity to provide this ES. Therefore, for the purposes of the ES catalogue of Slovakia, the calculation of ES water flow regulation was carried out as the landscape's capacity to provide this ES, which is a kind of *prevention* against the possible emergence of undesirable phenomena.

The map was compiled on the basis of available relevant documents (see Table 4.4 for a list of background maps). It represents a combination of two basic factors – the favourability of *local conditions* in terms of runoff regulation represented by the quantity and quality of the vegetation cover and soils (the expression of the so-called CN-curve based on vegetation and soil data) and the characteristics of the micro-basins in terms of transformation of runoff conditions (size, average slope, vegetation coverage). The result is provided in a form of the *relative scale of regulatory functions of the landscape and the micro-basins.*

The landscape's highest capacity to provide this ES is not typical for mountain areas but for larger valleys of watercourses, water reservoirs and lowland landscape with sufficient representation of forests or water elements (Fig. 4.15). The above-average protective capacity is shown by less rugged forested mountains, while low to very low capacity of the landscape is documented for deforested hills and rugged river basins with lower representation of forests. For a large part of the territory of

Table 4.4 Input data for capacity, demand and flow assessment of ES water flow regulation

Input data/ES	R4: water flow regulation
Capacity	Landscape use – CLS types
	Nature of vegetation – structure and quality of forest habitats (alternative C-factor)
	Soils – permeability classes
	Relief – average slope of micro-basins
	Structure of vegetation – the average value of CN-curve for micro-basin
Demand	Classification of micro-basins according to flow volumes/peak flood discharges
	Micro-basin classification according to flood risks
	Number of inhabitants of municipalities/areas in flood-threatened areas
	Definition of particularly sensitive areas – urbanized areas, residential and technical buildings, agricultural areas
Flow	The real effect of ecosystems – real protective effect according to micro-basins
	Degree of real ES action during real floods
	Number of residents within the reach of the ES – protection of citizens, prevention of financial losses

Fig. 4.15 Riparian forest and shore plants near the Danube (Patince) as an important element of water retention in the landscape, also providing flood protection function. (Author: D. Štefunková)

Slovakia, the average capacity of the landscape is typical for the regulation of runoff processes (Fig. 4.16).

Table 4.4 also includes useful indicators for determining the level of demand for this ES and its real use. Demand for ES can be assessed on the basis of the above-mentioned sources (flood risk maps, or classification by micro-basins). It is also

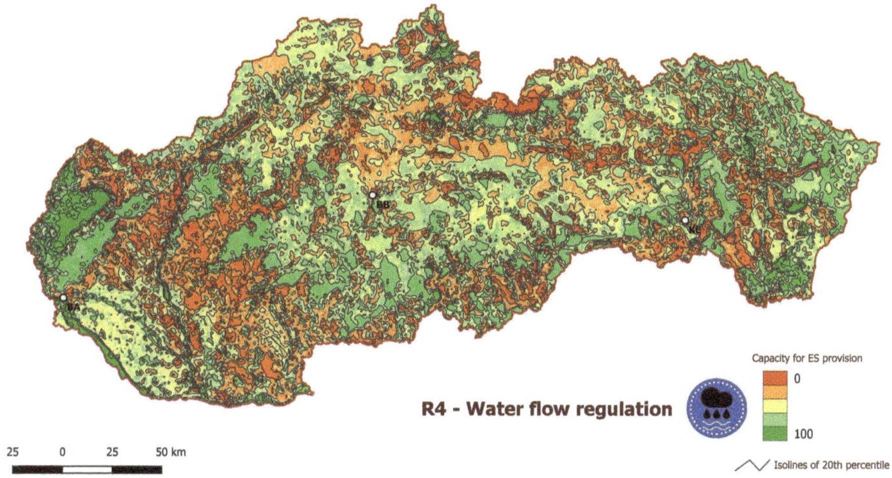

Fig. 4.16 Capacity of the landscape to provide ES water flow regulation

important to specify the number of inhabitants living in vulnerable territories and potential economic damage.

The real use of ES water flow regulation should be assessed on the basis of the real effect of ecosystems and the landscape during real or modelled floods – for example in the form of a flood-protected area, the number of protected residents or the value of avoided economic damage in territories with a real positive ES effect. As in the case of the previous ES, in the absence of the necessary data, proxy indicators may be used, for example, population density, spatial projection of settlements and other activities.

4.5 Local Climate Regulation (R5)

4.5.1 Definition and Brief Characteristics of ES

Local weather and climate are determined by the complex interaction of regional and global circulation characteristics with local topography, vegetation, as well as the configuration of water bodies (De Groot et al. 2002). According to Smith et al. (2012), ecosystems provide shelter from heat, UV radiation, wind and precipitation;

they regulate local temperature, the occurrence of droughts and the amount of precipitation.

Ecosystems regulate our climate at different levels. In cities and their surroundings, tree vegetation or urban forests provide shade during hot summer days and through evapotranspiration cool the surrounding environment, bringing benefits in terms of cost savings or reduced ozone production (Burkhard and Maes 2017). Evapotranspiration is the process of water intake by leaves, its conversion to water vapour and the consequent emission of water vapour into the atmosphere (Georgi and Dimitriou 2010). The conversion of water in leaves into water vapour cools the leaf and by releasing the vapour into the atmosphere through stomata also cools the surrounding microclimate (Hunter et al. 2012). Therefore, this physical phenomenon plays an important role in the water cycle and at the same time contributes to the provision of ES by vegetation.

First, providing shade with tree vegetation means changing the radiation balance which has two basic effects on humans. Plants capture part of the incident shortwave radiation, which increases the temperature on the Earth's surface, leaving the air temperature below the vegetation low. Second, reducing the effect of direct radiation on the human body reduces its physiological burden. These two effects of ES regulation of microclimatic conditions increase people's comfort during hot summer days (Ali-Toudert and Mayer 2007; Lee et al. 2013).

The shade provided by tree vegetation in cities has, in addition to temperature reduction, a positive effect on buildings – trees growing near buildings reduce their temperature during the summer days, thereby saving the cost of cooling them/air conditioning (Nakaohkubo and Hoyano 2011; Berry et al. 2013).

Based on the above definitions, the regulation of local climatic conditions can be characterized as the ability of ecosystems to regulate temperature and provide shade, to support the evapotranspiration process, to regulate the amount of incident solar radiation and to some extent regulate the spatial distribution of other microclimatic factors (e.g. wind, precipitation) and dampen the effects of some related processes (e.g. pollutants, dust, noise). In particular, these co-acting factors provide a local temperature reduction during days with high daily temperatures.

4.5.2 Methods Used to Assess and Identify ES

Biophysical methods in particular, but to some extent economic and sociocultural methods too, are used to assess this ES.

The basic and simplified assessment method (not only for this) of ES is the use of the so-called production matrix according to Burkhard et al. (2014), which expresses the relative potential, supply and demand for ES for the main types of ecosystems, or forms of landscape use. The regulation of local climatic conditions is provided to the highest extent by the index 5 forest and shrubby ecosystems, with the highest consumption and thus the deficit coming with index −5 built-up urban parts.

According to assessment work done by Pérez-Soba, Harrison et al. (2015) and Czúcz et al. (2018), the most used indicators of the assessment of this ES include, in addition to the characteristics of landscape use, in particular the climatic parameters (temperatures, precipitation, evapotranspiration, shading, wind, surface reflectance), spatial structure of vegetation (spatial distribution, cover, biomass volume), vegetation quality, representation and the nature of settlement vegetation (quantity, quality).

Local climate regulation has been assessed in national ES studies, for example in Germany and Romania. The indicators used for this EC included the volume of biomass, population density and the proportion of green areas in settlements (Germany) or meteorological data (temperatures, precipitation) and population distribution (Romania).

Based on the study of expert works, the following indicators, in particular, can be used for the biophysical assessment of the ES local climate regulation: temperature regulation, incident radiation regulation, shading and evapotranspiration. These indicators are mentioned in the vast majority of scientific papers in relation to the urban environment, in which they are easier to measure and assess, especially in relation to human health, and are more interesting because of the direct effect at the site of action. All four indicators of regulation of local climate conditions are interconnected and linked. In his work, Takács et al. (2014) support the findings made in recent decades that have shown air temperature reduction due to tree vegetation at the local level, especially during the day. The average air temperature below the tree canopy was, on average, 1–4 °C lower compared to the ambient air temperature. Hunter et al. (2012) state that the *tree canopy* can reflect, absorb or transmit incoming solar radiation depending on the type of vegetation, stands density, woody plants size, etc. The transmission of solar radiation through the *tree cover* in the summertime ranges from 4% to 30% and in winter from 40% to 80% (Shashua-Bar et al. 2010; Konarska et al. 2013).

From the point of view of ES provision, the quantifiable vegetation attribute, *leaf area index* (LEA), used in its assessment (Lee and Park 2008; Georgi and Dimitriou 2010) is important. Software tools such as FAPAR (Fraction of Absorbed Photosynthetically Active Radiation) can be used to assess the leaf area, using the current Landsat satellite images or Copernicus data.

Sociocultural methods can also be used for the purpose of assessing this ES – for example the contingent valuation method, which involves a direct determination of people's willingness to pay or accept compensation for a change in ES within a hypothetical market (Farber et al. 2006). Identifying the diversity of views on the well-being based on ES cultural value (Fernando et al. 2013) has shown that people living in the countryside combine their well-being with provisioning ES (food, cattle, fishing) and, conversely, people living in cities prefer (value highly) regulatory ES, in particular regulation of microclimate and air regulation. Promoting green infrastructure was an important part of the ES's assessment in Italy (Capotorti et al. 2015).

From the economical methods, the following methods are suitable, in particular: cost-saving methods (which would arise in case of failure of the given ES – e.g. costs of air conditioning or heating, etc.) and method of benefit transfer from other territories.

Fig. 4.17 Wetland habitat contributes significantly to the local climate regulation of – Šúr Site of Community Importance. (Author: J. Černecký)

4.5.3 The Main Types of Landscape and Ecosystems Which Provide ES

Based on analysed works and assessment of ecosystem capacity and landscape of Slovakia, the following may be considered as the main types of landscapes/ecosystems regulating the local climate: forests and other elements of permanent vegetation, wetlands (peat bogs, marshes and other wetlands – Fig. 4.17), water bodies, watercourses and shore vegetation, to a limited extent also grasslands – meadows and pastures and subalpine and alpine communities.

The regulation of local climatic conditions within Slovakia is mainly provided by forest ecosystems, to a lesser extent by non-forest communities and potentially also by agricultural land. To assess microclimatic conditions, it is necessary to consider the quantitative representation of individual types of ecosystems in Slovakia, not only their quality.

In terms of regulation of the local climate, forest ecosystems also dominate this ES, both in terms of quality of provision and quantitative representation. The area of grassland and herbal habitats plays an important role because, due to its significant presence, it can be considered as the second most important category of ecosystems after forest ecosystems after considering consumption/demand. If only the quality of provided ES is taken into account, peat bogs, marshes and raised bogs are also important. Other ecosystems are less involved in the creation of this ES.

On the contrary, demand for this ES is significantly higher than production in built-up areas, especially in residential areas.

4.5.4 Importance of ES in Terms of Nature and Landscape Protection in Slovakia

The importance of nature and landscape protection for the regulation of the local climate is quite notable. The prerequisite for ES provision is the condition and dynamics of ecosystems, the nutrient cycle and the connection with other ecosystems. These conditions can be ensured by protecting them, maintaining a favourable ecosystem status and managing them in protected areas (Fig. 4.18). Protected areas most often include large-scale continuous forest ecosystems, which are key to regulation and co-creation of the local climate at the national level.

Green elements in urbanized environments improve the environment, i.e. they increase human comfort through the provision of several ES, but mainly by local climate regulation. With the current negative trend of increasing temperature extremes in the summer months, caused by climate change and resulting weather extremes, the regulation of local climatic conditions is very important, as well as an easily identified function of ecosystems by the general population. An example of this is the seeking of shade during hot days under the tree canopy which, through physical phenomena, reduce temperature, reduce incident radiation and cool the air in the surrounding environment. Paradoxically, landlocked countries like Slovakia are hit by increasing average annual temperatures the most. This is evidenced by measurements in recent years when the Czech Republic and Slovakia have recorded some of the highest increases in average annual temperatures among all EU Member States.

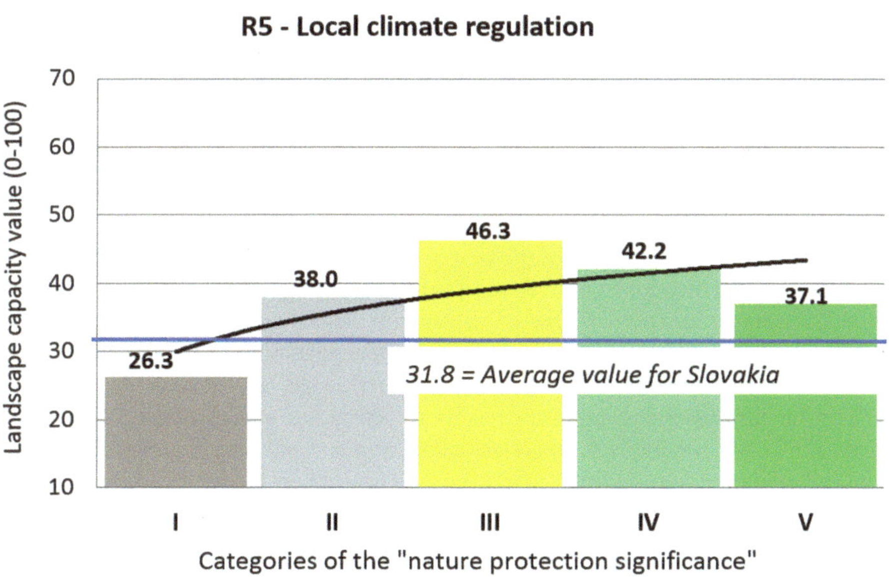

Fig. 4.18 The relationship between ecosystem service R5 and the significance of the territory of Slovakia in terms of nature and landscape protection

The creation, protection and maintenance of permanent vegetation elements in towns and villages in Slovakia, such as city parks, forest parks, orchards, tree alleys, gardens and woody plants planted in housing estates, together with water elements, represent elements of the concept of *green infrastructure* and TSES. The aim of these is to interconnect the natural/semi-natural areas or ecosystems in the urban environment. The importance of building green infrastructure elements is obvious and justified, so this measure is part of the updated Adaptation Strategy of SR to climate change 2018.

In urban areas, it is necessary to support the provision of ES by establishment and maintenance of a wide range of woody and herbaceous vegetation in parks and forest parks, preventing any harvesting of woody plants or tree alleys along roads, watercourses and housing estates, since uniform and *sterile* semi-natural ecosystems do not provide ES in full scale and degrade over time.

Nature protection objectives often focus on achieving a favourable state of habitats and species located in protected areas. However, the implementation of the measures to improve/maintain the status does not only have an effect on the subject of protection but also the provision of an accompanying ES, which is essentially a contribution not only for nature protection but also for residents in the form of improved local climatic conditions.

4.5.5 ES Assessment for the Territory of Slovakia

Ecosystem functions aimed at regulation of the local climate are inherently local and therefore often not understood and assessed at the national level. In addition, its assessment often uses parameters similar to those of the ES global climate regulation, and it can be stated that these two ES also significantly intersect in the assessment with regard to the types of ecosystems providing the service and also with regard to the assessment of potential, provision and demand.

The assessment of the territory of Slovakia in terms of the potential or provision of this ES is not implemented, although the key role of vegetation and especially of forests is evident. The importance of the so-called atmospheric (or climatic) functions of the forest is also mentioned by Čaboun et al. (2010) or in other assessments of non-production forest functions. Climatic functions of vegetation in urban, especially city environment (Fig. 4.19), are investigated by, for example, Supuka (1998). However, a comprehensive assessment of the territory of Slovakia has not yet been prepared.

For the pilot assessment of the capacity of Slovakia's territory from the point of view of this ES (see Table 4.5), we used data on the regulatory function of vegetation (based on the state of forest stands and the use of non-forest areas), which were also used in the R1 regulatory service. The basic classification of the area was subsequently refined with the use of two indicators – the coefficient of climatic conditions (temperature ratios, amount of solar radiation) and the vegetation efficiency coefficient (based on the combination of indicators NDVI (normalized difference vegetation index) and FAPAR). These two indicators were obtained from the database of the European system RS Copernicus (available online: www.copernicus.eu/en).

Fig. 4.19 City parks provide regulation of climatic conditions in cities. (Author: J. Černecký)

Table 4.5 Input data for capacity, demand and flow assessment of ES local climate regulation

Input data/ ES	R5: local climate regulation
Capacity	Map of current landscape structure – reclassification as appropriate for ES provision
	Species composition and structure of forest stands (classification, types of stands, age of stands)
	Climatic data – global solar radiation and avg. temperature of the growing season
	FAPAR (Fraction of Absorbed Photosynthetically Active Radiation – an indicator of photosynthetic activity of vegetation)
	NDVI (normalized difference vegetation index)
Demand	Climatic classification of the territory – areas with the highest temperatures and sunlight, insufficiently provided with moisture
	Classification by population of municipality/region
	Special areas of demand – residential areas, city centres, recreational areas,
Flow	Real effect of vegetation – improvement of local climate parameters (temperatures, radiation, shade, air humidity)
	Number of residents within reach of ES

Both indicators implicitly express the representation of vegetation and photosynthetic activity (NDVI, FAPAR).

The calculated data was reclassified in a similar way as in the case of the other ES into the scale of the landscape's relative capacity to provide ES local climate regulation. The assessment result is shown in Fig. 4.20.

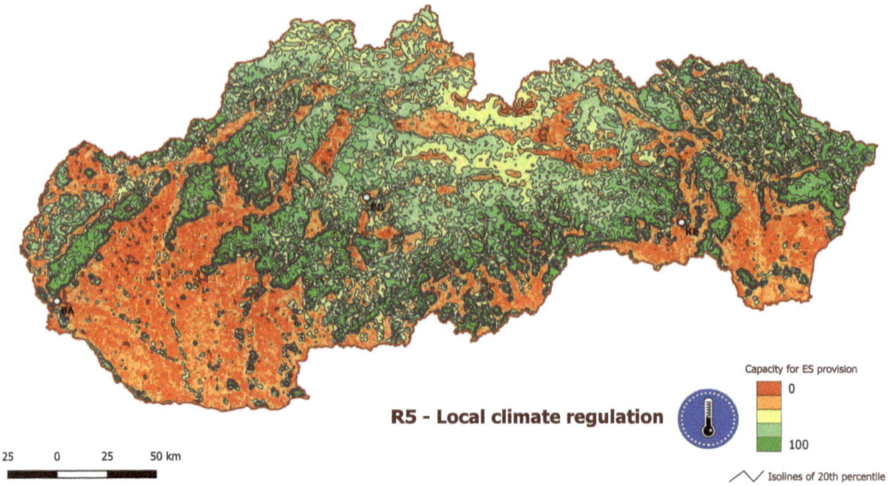

Fig. 4.20 Capacity of the landscape to provide ES local climate regulation

The best habitats in terms of provision of regulation of local climatic conditions are natural forest stands and forests in good condition (species composition, age structure) – however, the quality of such high-quality forests is decreasing in Slovakia, as evidenced by the value of satellite vegetation indices. To a lesser extent, this ES is provided by non-forest ecosystems – a mosaic agricultural landscape with sufficient representation of permanent vegetation, watercourses and areas, grasslands and in some cases orchards. Intensively used agricultural land, especially arable land, is hardly involved in the provision of this ES. In urbanized areas, the role of the settlement vegetation, especially of larger parks, is irreplaceable – at the national level; however, their significance is almost unregistered due to their small size.

Table 4.5 also includes the basic indicators which can be used for future expression of the level of ES demand as well as its real use. As is the case with air quality, the highest demand for this ES is in built-up areas – a suitable indicator comes in the form of the number of inhabitants living in a particular territory. Although city parks, tree alleys or gardens and orchards are involved in the provision of this ES within cities, the demand for this ES largely exceeds its provision. Demand can be expressed also by the need for regulation of microclimate (definition of exposed areas in terms of temperature and solar radiation) or the existence of special types of territories requiring this method of regulation (residential areas, facility areas in cities, etc.).

The ES flow is conditioned by its real use – i.e. the degree of improvement of climate conditions due to ecosystems, the number of inhabitants living in the affected area and so on. Obtaining such data at the national level can be problematic, so proxy indicators can be used – for example population density, the spatial extent of settlement, overall values of solar radiation, extreme summer temperatures and the like.

4.6 Global Climate Regulation/Carbon Sequestration (R6)

4.6.1 Definition and Brief Characteristics of ES

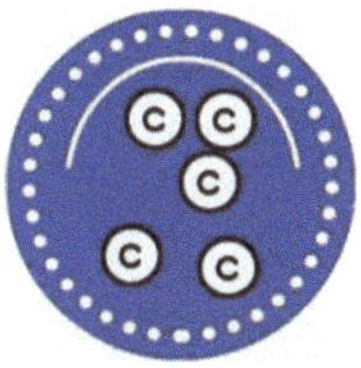

The main role of global climate regulation through ecosystems is to maintain liveable climate and thus maintain a favourable chemical and physical composition of the atmosphere for human beings. Natural forest ecosystems and wetland ecosystems as well as maritime/coastal areas maintain suitable atmospheric conditions for life on Earth and regulate climate at global level (Maes et al. 2015). The biodiversity information system in Europe (BISE 2019) identifies global climate regulation as one of the most important ES on a global and European level, as European inland ecosystems represent a stock of 7–12% of *pure* carbon from anthropogenically produced carbon emissions, according to measurements from 1995.

ES experts and other authors agree in their publications on the basic functions or on the primary indicators which contribute to the production of this ES or support global climate regulation partially. Mooney et al. (2009) state that ES global climate regulation is mainly aimed at the issues of greenhouse gases, so the carbon storage, carbon sequestration and greenhouse gas regulation are the primary indicators of this service. However, secondary indicators play an important role in the global climate regulation, such as above-ground and underground biomass, landcover, carbon deposited in soil, nutrient flow and soil characteristics. Burkhard and Maes (2018) identify the basic and supporting ecosystem functions to maintain the global climate regulation service. Primary functions include *net primary production, carbon storage* and *conservation carbon stock.* Supporting ecosystem functions are defined by the regulation of ecosystem dynamics, ecosystem stabilization processes, ecosystem resilience, the development of complex ecological networks and the development of ecosystem diversity/habitat quality.

Carbon sequestration is a natural process which significantly contributes to climate regulation by capturing and long-term storage of atmospheric CO_2 in the soil (CO_2 being the major greenhouse gas) (Luyssaert et al. 2007). Carbon sequestration involves the transfer of atmospheric CO_2 into long-life *reserves*, i.e. carbon stock and its safe storage, so it does not immediately return to the cycle (Lal 2004). Pure primary production represents the net amount of carbon assimilated by green plants/vegetation (within a given time period).

The total land-related organic carbon reserves (in soil and vegetation) are estimated at 3500 Pg C, and most of it (up to 75%) is stored in the soil. It is almost five times the amount than the amount of carbon in the atmosphere. The carbon deposited in the soil comes mainly from dead organic material. The main factors influencing the state of soil organic carbon reserves are the landcover consisting of inland ecosystems and their habitats, land management and local climatic conditions. Land use change and management practices can lead to carbon flow imbalance (Burkhard and Maes 2017).

Increasing the area of wooded land in the UK has contributed to improving climate regulation through higher carbon sequestration while improving ES associated with timber production. The projected changes in emissions (within the business-as-usual scenario) resulting from land use and forestry changes in the next 10 years will change the net carbon stock to the source of its production. The effects of failure to provide this ES would be particularly pronounced in urban areas and would make the climate stress worse for a large number of people (UK NEA 2011).

In summary, according to MEA (2005) – global climate regulation is the *final ES* which provides climate regulation through biogeochemical and biophysical processes in such a way as to avoid adverse effects on humanity and biodiversity.

4.6.2 Methods Used to Assess and Identify ES

Global climate regulation such as ES is often indicated by carbon sequestration or net primary production, probably as a result of the great attention paid to climate change (Maes et al. 2015). Net primary production is the basis of this ES but also of many other ES and is, therefore, the most frequently mapped indicator (Burkhard and Maes 2017).

Biophysical methods for assessing this ES are based on soil *carbon pools*. This indicator mainly affects the process of sequestration and net primary production as a potential for carbon pool creation (Haberl et al. 2007). In the framework of biophysical methods of ES assessment, the InVEST and ARIES modelling tools need to be highlighted. Both models work in ArcGIS environment and are freely available. The primary input to these models includes the land cover and land use maps, complemented by the socio-economic and ecological parameters (soil carbon pools, average annual precipitation).

A study from Northern Germany (Maes et al. 2018) for the assessment of this ES applied quantitative indicators derived from Corine Land Cover categories such as annual gross primary production, net primary production, soil organic carbon and carbon pool compared to qualitative indicators.

Another way of ES assessment includes the use of the production matrix. For example, according to Burkhard (2014), each ecosystem-provided ES is rated on a scale of 1–5 (low to very high benefit), and ES consumption is rated on a scale from −1 to −5 (low to very high demand). Value 0 is attributed to services and ecosystems which do not produce or consume the ecosystem service. Burkhard's production matrix index values show that wetland habitats and forest habitats have the highest provision index for ES global climate regulation. The demand/consumption index is the largest in cities and densely populated areas.

Cascade model for the assessment of ES global climate regulation was used in the national ES assessment in Finland (Jäpinen and Heliölä 2015). The following indicators were assessed: (1) habitats with carbon pools, forests, wetlands, inland water bodies, farms and urban areas (structure); (2) carbon balance, sequestration rate (function); (3) climate regulation and stabilization (benefit); and (4) avoided costs of negative climatic consequences, actual/used value of stable climatic conditions (value). This ES was similarly assessed in Luxembourg (Kleeschulte and Ruf 2016), where they used the capacity indicator (modelled carbon pools per mapped unit), the ES balance/flow indicator (carbon storage per mapped unit) and the ES benefit indicator – carbon sequestration value in dollars per tonne.

According to Frélichová et al. (2014), the carbon sequestration or carbon pools represent the biophysical method for assessing/valuing global climate regulation. Much more options for the assessment of this ES come from *economic methods*: avoided cost, benefit transfer, contingent valuation, emissions trading scheme, marginal abatement cost, direct market valuation and the social cost of carbon. The average economic value of ES climate regulation for the Czech Republic according to this study was set at EUR 4015.78/ha.

4.6.3 The Main Types of Landscape and Ecosystems Which Provide ES

Based on the above-presented approaches and methods of ES assessment and identification and in accordance with the production matrix by Burkhard (2014), the following can be considered the main types of landscape/ecosystem which provide global climate regulation: forests and other wooded landscape; peat bogs, marshes and other wetlands; meadows and pastures and alpine vegetation, subalpine shrubs, raised bogs and inland surface waters and riparian vegetation.

Forest ecosystems cover a large part of Europe, and their share in global climate regulation is, therefore, most prominent. Trees and other woody vegetation process and store large amounts of carbon through their assimilation organs. Larger reserves of organic carbon are further produced only by peat bogs. Meadows and pastures, alpine vegetation and riparian vegetation contribute to the ES supply at a lower level in terms of area and vegetation but are important in terms of quantity.

In terms of preliminary analyses of the provision of ES, habitats with a large area are more significant in Slovakia, as they provide the ES on a considerable area, as opposed to habitats, which are the most significant in terms of the quality of ES provision, but their area is negligible. In terms of both quality and quantity, the provision of this ES is dominated by forest ecosystems, which also have a high potential, as well as the value of the provision of this ES and, at the same time, high quantity. Peatbogs are included in the category of high quality. In case of arable land ecosystem, when taking into account the area within Slovakia, consumption of this ES is expected to be significantly higher than its production by this ecosystem. The built-up area also does not produce this ES, while ES consumption is obvious and to a high degree.

4.6.4 Importance of ES in Terms of Nature and Landscape Protection in Slovakia

Ecosystems are highly involved in carbon sequestration and storage, as well as in the production of new biomass as key indicators of global climate regulation. Carbon sequestration by above-ground biomass and its storage in soil reduce the rate of CO_2 increase in the atmosphere, which, along with other greenhouse gases, affects the processes of global warming and climate change on Earth. Based on this, it can be stated that ecosystems, especially ecosystems in a favourable state, mitigate the effects of global warming on biodiversity, such as increasing average annual temperature, shifting of vegetation belts, wind calamities and related calamities caused by bark beetle insects, drying out of aquatic and wetland habitats and decreasing level of groundwater, shorter periods of permanent snow cover in the mountains, spread of invasive and expansive species, etc.

In Slovakia, most importance of the carbon sequestration and net primary production is attributed to forest ecosystems (more than 38% of the SR area). For the practical protection and conservation of forest habitats in Slovakia, several national parks, protected areas, nature reserves and special areas of conservation were declared. Particular attention should be paid to the protection of Natura 2000 areas under which forest habitats of European importance are protected – including, for example, NNR and SAC Svrčinník (Fig. 4.21). In terms of quantity, it is the large-scale protected areas which have the greatest benefit, namely, the most widely represented habitats in them – Ls5.1 beech and fir-beech forests and Lk1 lowland and submontane hay meadows. Despite the relatively common occurrence within Slovakia, the habitats just mentioned playing a key role in maintaining and keeping global climate regulation. The continuous large-scale areas of the Ls5.1 and Lk1 habitats are the most important in terms of the provision of this ES and are mainly located in the national parks of Slovakia.

In case of this ES, it is also necessary to emphasize the value and benefit of primaeval forests and primaeval forest remains, which represent a prime example of the maximum benefit of global climate regulation and are among the best carbon pools of all the ecosystems. The most qualitatively significant carbon pools in the form of deposited organic residues in Slovakia are provided by peatbog habitats, for the protection, of which several small-scale protected areas such as NR Rojkovské rašelinisko peatbog, NNR Rakšianske rašelinisko peatbog (Fig. 4.23) and others have been declared. The positive relationship of this ES and the significance of the territory of Slovakia in terms of nature and landscape protection is also apparent from Fig. 4.22 (Fig. 4.23).

Seeing the protected areas as a basic tool for the protection of biodiversity in Slovakia can be enriched by one of the most significant benefits provided by these areas with the application of the ES concept. Therefore, in economic terms, consumers in protected areas are no longer represented by only the habitats and species but also by people as one of the main consumers. In this case, this is associated with ES essential for survival and key for the adaptation to the current and incoming climate change. Changing the view of the nature and benefits of protected areas is

Fig. 4.21 Primeval forest remains of natural spruces in NNR Svrčinník represent a carbon pool. (Author: J. Černecký)

essential and very important in terms of future scenarios for the development of not only biodiversity but also the survival and quality of life of humans themselves.

4.6.5 ES Assessment for the Territory of Slovakia

The assessment of the ES global climate regulation/carbon sequestration by a market prices method was described by Považan et al. (2014b). They specifically describe two ways of calculating carbon stock in both above- and below-ground biomass and underline the need to develop clearer benchmarks for this indicator for different types of forest ecosystems and recommend taking into account the impact of climate change on carbon storage. The peatbogs are the most important carbon pools, but their massive drainage during the collectivization in Slovakia caused their degradation and vanishing of habitats. Restoring and protecting peatbogs is key to

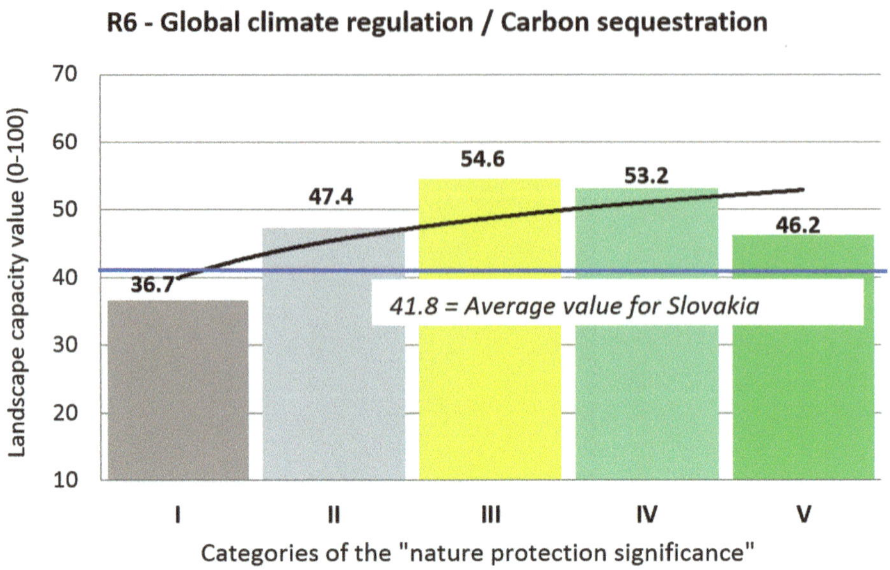

Fig. 4.22 The relationship between ecosystem service R6 and the significance of the territory of Slovakia in terms of nature and landscape protection

mitigating climate change; even though they are considered small-scale habitats, they are important in terms of the quality of provision of this ES at the local level.

ES assessment through biophysical indicator – a measurement of organic carbon in the surface layer of the soil – is investigated in Slovakia by Skalský et al. (2017). Carbon stocks in agricultural soils in Slovakia can be estimated on the basis of NPPC-SSCRI data. In the past, a map of organic matter content in soils of the SR was prepared (Bielek in Granec et al. 1999).

Despite these approaches, the comprehensive assessment of the ES global climate regulation/carbon sequestration in Slovakia is not performed. The assessment should be based on three aspects: capacity, demand and real production/consumption of this service. As mentioned above, *forest and selected non-forest ecosystems are important in terms of ES provisioning capacity*. The *peatbogs* have the highest quality for provision of this ES, but their area is very small to fundamentally affect the overall value at the national level. The need to protect them is that much greater. Therefore, as with most other regulatory services, forest ecosystems must be given the greatest importance.

For the pilot assessment of the Slovak territory's capacity, we also used the data on the regulatory function of vegetation assessed under the R1 regulatory service as the basis for this ES. The coefficient used to refine this value was the FAPAR indicator expressing the rate of photosynthetic activity of vegetation and was obtained from the source RS Copernicus (available online: www.copernicus.eu/en). The second aspect of the provision of this ES (carbon retention rate in soils) was expressed by the capacity of the soil to accumulate carbon, based on the organic matter content

Fig. 4.23 In terms of quality, peatbog retains the most carbon compared to other types of habitats (National Nature Reserve Rakšianske rašelinisko). (Author: J. Černecký)

in the soil subtypes and the depth of the soil cover. The overall capacity of vegetation and soils for carbon capture and storage was then expressed in a relative scale of landscape capacity to provide this ES – the assessment result is shown in Fig. 4.24.

As is the case with the previous ES, the best quality habitats for the provision of this ES are forest areas of good quality (species composition, age structure), but the non-forest ecosystems are also important – already mentioned *wetlands* (with a very low occurrence), production meadows and pastures – significant carbon supply is also saved in the top quality agricultural soils (deep soils with good-quality humus layer and high organic content).

Table 4.6 also shows the basic indicators which can be used for future expression of the level of ES demand as well as its real use. In this case, it is not easy to establish demand indicators – it could even be said that the need for global climate regulation is the same throughout Slovakia. However, if we want to distinguish some areas, then densely built-up areas and places of consumption can be rightly considered to be places of increased demand – i.e. intensively used agricultural areas.

From the point of view of the real production/flow of this ES, it is clear that there are much more ecosystems with only the average value of the provision of global climate regulation than the potential. In order to increase the provision and quality

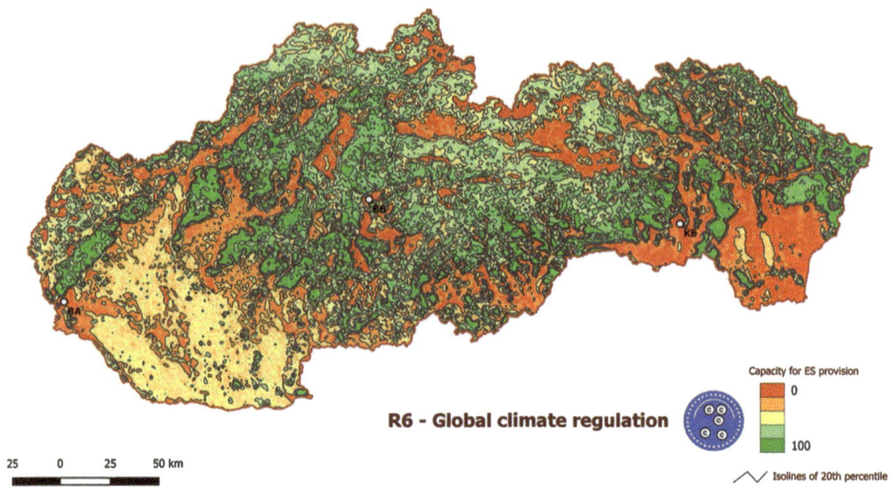

Fig. 4.24 Capacity of the landscape to provide ES global climate regulation

Table 4.6 Input data for capacity, demand and flow assessment of ES global climate regulation/carbon sequestration

Input data/ ES	R6: global climate regulation
Capacity	Map of current landscape structure – reclassification as appropriate for ES provision
	Species composition and structure of forest areas (classification, forest types, forest age)
	FAPAR (Fraction of Absorbed Photosynthetically Active Radiation – an indicator of photosynthetic activity of vegetation)
	Soil classification based on organic matter content and soil depth
Demand	Special areas of demand – residential areas, places of carbon consumption (agricultural land)
	Classification by population of municipality/region
Flow	Real ES provision – carbon storage in vegetation and soils
	Number of residents within the real effect of ES

of this ES, it is necessary to improve the status of watercourses, reduce the size and intensity of forest interventions, increase the age of the forests and substantially protect peatbog and wetlands. The last should be done in places where they are already protected and try to revitalize the wetlands in places from which they vanished, as their size in relation to other ecosystems is extremely small. In agricultural areas, it is also appropriate to limit deep ploughing, which contributes to the release of carbon from the soils.

4.7 Biodiversity Promotion (R7)

4.7.1 Definition and Brief Characteristics of ES

Biodiversity expresses the diversity and variability of living organisms and ecosystems. Different organisms, species and communities differ in their properties and functional characteristics, as well as in the share within ecological processes. Species and ecosystem diversity promotion as the ES is seen as a result of complex interactions between biotic and abiotic environmental components, which support species life cycles. Among these are the conservation of habitats and species, the preservation of key habitats for animal husbandry, as well as the conservation of genetic diversity, and the promotion of cultivated and farmed species in nurseries, arboretums, breeding ponds, etc. It is very difficult to accurately describe the importance of all species biodiversity for humans. Approximately 40% of the global economy is estimated to be based on biological products and biological processes. Biodiversity promotion was initially specified in the ES classification as a separate supporting service group (Kumar 2010; MEA 2005); in other classifications, it has already been included in regulatory services (Haines-Young and Potschin 2013).

The main importance of the ES *biodiversity promotion* is to ensure the proper functioning of ecosystems, which also affects the provision of other major services (Becerra-Jurado et al. 2016). Ecosystems themselves contribute to biodiversity promotion by providing living space and refuges to different species of plants and animals; by providing them with food and shelter opportunities, space for plant and animal reproduction, space for migration or spreading within the landscape (seed dispersal by insects, birds and other animals) and biotopes for pollinators; and by participating in the nutrient cycle and the like. In fact, with a few exceptions, we could consider most of the ES to be the benefits gained from maintaining and promoting biodiversity. Maintaining the diversity of nature as a whole – especially the number of plant and animal species, their regional and local populations and genetically modified variants – is one of the basic tasks of not only for nature conservation, science and culture but also for economic activities (Čaboun et al. 2010).

Higher biodiversity increases the potential of terrestrial, freshwater and marine ecosystems to provide different benefits to society, such as soil formation, pollination, erosion and other natural hazards regulation, regulation of air and water quality or provision of materials, as well as space for education, inspiration, or physical use of nature and landscape. Higher biodiversity promotes the functioning of all ecosystems and also contributes to maintaining ecological stability (see Fig. 4.25).

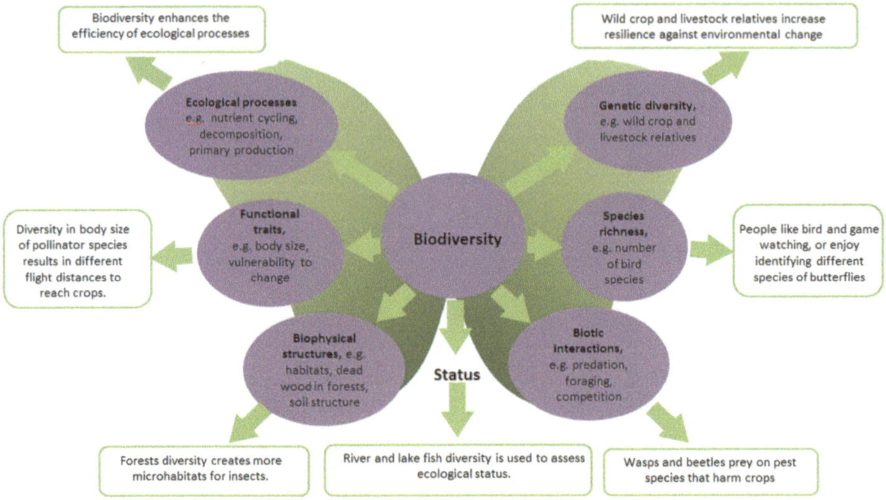

Fig. 4.25 The diverse role of biodiversity in promoting the provision of ecosystem services and assessing the ecosystem status. (Source: Maes et al. 2013)

4.7.2 Methods Used to Assess and Identify ES

In biodiversity promotion, the need to protect and preserve the biodiversity of individual species and habitats is of interest to science research since its inception. The species and ecosystem diversity itself can be expressed using different diversity indices (Jurko 1990; Loh and Harmon 2005; Pielou 1975). The ecological status of the landscape and the importance of the bio-component in the landscape are characterized by, for example, *ecosozological characteristics of the gene pool* (a rarity, endangerment, endemites, protected plants), *degree of ecological stability threats* and so on (Barančok and Barančoková 2015; Halada et al. 2011; Špulerová 2007). The assessment of this ES was preceded by the theory of ecosystem functions, which began to develop more intensively in the second half of the twentieth century. According to Kontriš (1978), the vegetation function is the highest category expressing the aggregate real or potential use of the effects of vegetation, which participates in the creation of ecosystems and the creation of ecological conditions of the environment, aimed at meeting the economic and social needs of society. De Groot (1992) defines ecosystem functions as *the ability of natural processes and components to deliver goods and services which directly or indirectly satisfy people's needs*. Mapping of habitats and their characteristics (such as functional properties, ecosystem structure) is a determining indicator for ES assessment (Lavorel et al. 2011).

The capacity of current ecosystems to support species and ecosystem diversity has been assessed using a variety of methods, most commonly biophysical assessment, participatory methods and economic expression of ecosystem value.

Biophysical assessment was mostly based on habitat mapping, the determination of ecosystem basic state and proper indicator selection. Such an assessment has been used in several national ES assessments:

- In Flanders, on the basis of selected criteria (a rarity, biological quality, vulnerability and ecosystem resilience), five classes of ecosystem assessments were distinguished, from the built-up area (no value) to very valuable areas (Stevens et al. 2015).
- Biodiversity indicators were assessed relatively in detail on a five-degree scale as part of the national assessment in Bulgaria at different levels: for plant biodiversity (cover and type of vegetation layer – using aerial and satellite images, habitat type, number of protected species), animal diversity (number of protected species), habitat diversity (share of natural habitats, fragmentation of green infrastructure) and spread of invasive species (Vranic et al. 2016).
- For the ES national assessment in Ireland, the design of indicators was based on the international CICES classification and on the European ES assessment methodology (European Commission 2014b). The following indicators have been proposed as indicators for this ES: High Nature Value (HNV) areas and ecological status of aquatic ecosystems (Parker et al. 2016).
- A similar approach has been selected for the national ES assessment in Luxembourg, where capacity indicators have been proposed for this ES (capacity indicator – habitat quality, area for biodiversity support (European areas with HNV) – and balance indicators flow indicator – number of species, biodiversity indicators, weighted index of the Birds Directive per unit area) (Becerra-Jurado et al. 2016). A map of ecosystems with habitat values has been included in the assessment because the authors assume that only healthy ecosystems are capable of sustainably providing this type of service.
- The MAES methodology was also applied to the national ES assessment in Italy – the following were proposed as indicators for the assessment: ecosystem status, degree of naturalness/hemerobia, nature conservation status, difference between real and potential vegetation, fragmentation of ecosystems and limiting indicators for achieving favourable habitat status (Capotorti et al. 2015).
- Indicator design in Finland was based on a cascade model, and important ES were assessed from four different aspects: structural (habitat area and status), functional (shelter and food possibilities, measured by reproduction success), utility (population vitality) and value (cost savings for revitalization and other management measures) (Jäppinen and Heliölä 2015).
- Practical assessment of forest quality at the landscape level is investigated by authors Dudley et al. (2012), who proposed the presence of rare and endangered species as the indicator for the *biodiversity and genetic resource protection* service.

Participative methods for ES biodiversity promotion were used by several authors. Burkhard et al. (2012, 2014) developed a matrix for the Corine Land Cover (CLC) categories and 29 ES grouped into four basic categories, based on MEA (2005). Based on expert estimates, they set the capacity to provide the ES on a five-degree scale (from no relevant capacity to very high relevant capacity). A similar assessment tool has been applied in several case studies based on expert estimation and public participation (Bezák and Bezáková 2014; Vihervaara et al. 2010). Expert assessment, combined with various spatial data (analytical maps) grouped into themes (instead of exclusively using landscape structure data), was used in the *GreenFrame* method developed in Finland to determine the ES capacity, especially for green infrastructure (Kopperoinen et al. 2014).

Another option for ES assessment is the visual modelling of scenarios, based on predicting the landscape's development in case of certain pressures affecting the landscape. As part of the behavioural research, these scenarios can be subsequently assessed and commented on by local stakeholders in order to select an optimal model of development of the assessed territory. This approach was applied to the national ES assessment in Denmark, using the following indicators of biodiversity change for seven assessed scenarios: coverage of landscapes important for the protection of rare species, habitat continuity and their structure. Three scenarios were aimed at promoting species and ecosystem diversity (Termansen et al. 2017).

Economic approaches present another option for assessing this ES. The ES monetary value based on their ecological value has been investigated in the Czech Republic (Seják et al. 2010). The ecological value of natural and semi-natural habitats mapped within the NATURA2000 system was calculated on the basis of expert scoring according to eight defined criteria (Seják and Dejmal 2003). Subsequently, the authors derived the initial monetary value by analysing the effectiveness of actual revitalization measures. In Finland, researchers also tried to express the annual value of forest ecosystems based on the assessment of the loss of biodiversity, expressed by the need to create habitats for 650 endangered species (Matero and Saastamoinen 2007).

4.7.3 The Main Types of Landscape and Ecosystems Which Provide ES

Ecosystems providing ES biodiversity promotion in Slovakia can be assessed from two points of view. From the point of view of quantity, the most common are forest habitats which cover about 40% of the country's territory – but they are altered to varying degrees by humans, which negatively affects the capacity to provide ES R7. The best-preserved good-quality habitats provide this ES in full, but their area is often negligible from a national perspective.

In terms of conservation of species and ecosystem diversity, the most endangered and rare habitats deserve the greatest attention, including calcareous marshes with great fen-sedge and *Caricion davallianae* species, oligotrophic to mesotrophic waters with benthic vegetation of *Charophyta*, active and degraded raised bogs with natural regeneration (Fig. 4.26) and, in general, biotopes associated with sands, peatbogs, alpine environment and xerothermic habitats. In addition, attention should

Fig. 4.26 Active raised bogs are very rare in Slovakia, they are a habitat of many rare and endangered species – Orava region, Rudné – Suchá Hora. (Author: J. Špulerová)

be paid to other European and national habitats which can contribute to biodiversity conservation (Černecký et al. 2020).

In terms of the assessment of this ES, it is important to distinguish habitats and the degree of their naturalness (natural, semi-natural and anthropogenic) as well as taking into account the status of habitats and species of European importance based on the EU Habitats Directive (Article 17) and the Birds Directive (Article 12). The assessment results are available online at www.biomonitoring.sk. The distinction and detailed description of plant species typical for individual habitat categories are contained in the Catalogue of Habitats of Slovakia (Stanová and Valachovič 2002). Other publications describing and assessing the status of habitats and species include, for example, Monitoring of Plants and Habitats of European Importance in the Slovak Republic (Šefferová Stanová and Galvánková 2015) and Monitoring of Animals of European Importance in the SR (Janák et al. 2015).

Other anthropogenic ecosystems which create elements of green infrastructure and create habitats for many animal species, thus contributing to biodiversity promotion, are important mainly in urbanized or intensively used agricultural landscape.

4.7.4 Importance of ES in Terms of Nature and Landscape Protection in Slovakia

The main benefit of this ES is the improvement of the conditions for maintaining the gene pool of plants and animals, creating suitable habitats, proper food and shelter opportunities for species migration, which is in line with the interests of nature and landscape protection in Slovakia. This ES, therefore, promotes the protection of nature and landscape the most prominent as it is directly aimed at promoting species and ecosystem diversity. This fact is clearly evident from Fig. 4.27, which shows the relationship between the landscape's capacity for provision of this ES and the significance of the territory in terms of nature and landscape protection.

From the viewpoint of ecosystem and species diversity protection as well as ecological stability and variability of the whole landscape of Slovakia, the most important tool is the existing network of protected areas (national network of protected areas, areas belonging to the European system of protected areas NATURA 2000 and internationally important areas identified under various international conventions) as well as biocentres and biocorridors of ecological networks from local to national level. The subject of protection of protected areas is precisely to enable/maintain the natural development of ecosystems as such; with respecting the values created by traditional forms of farming, the result of which comes in the form of rare communities of established non-forest habitats contributing to increased biodiversity. The established system of protected areas provides a precondition for promoting ecosystem stability at the national level. However, the mere fact that a

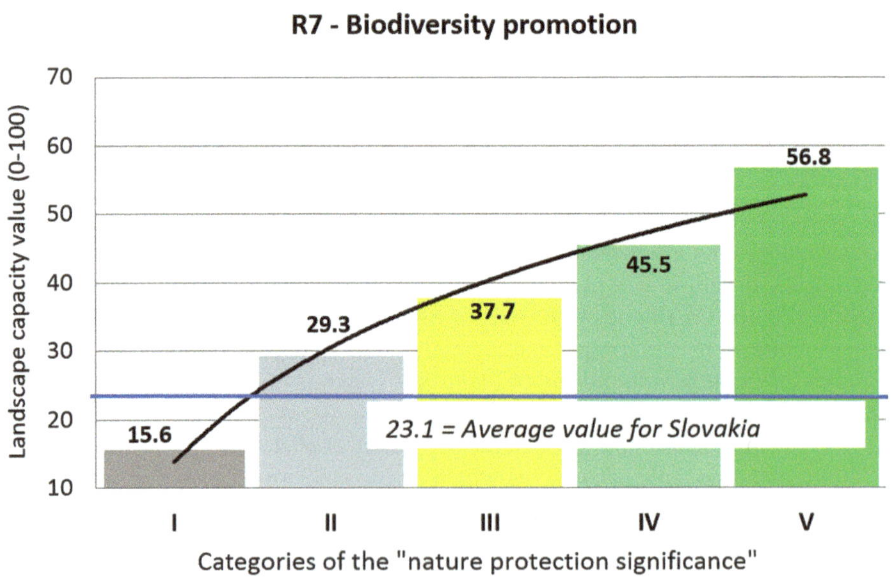

Fig. 4.27 The relationship between ecosystem service R7 and the significance of the territory of Slovakia in terms of nature and landscape protection

territory is declared protected will not prevent the continuing trend of biodiversity loss (Schröter et al. 2019). Therefore, it is important that individual protected areas have prepared and approved management programs and that individual measures to revitalize and maintain a favourable habitat and species status are implemented in practice.

Elements of the territorial system of ecological stability (ecological networks) are not covered by a special level of protection unless they are part of protected areas, but their creation and maintenance are of public interest. Act no. 543/2002 on Nature and Landscape Protection (National Council of the Slovak Republic 2002) in its Article 3 par. 3 states that, where entrepreneurs and legal persons intend to carry out activities in the landscape which may threaten or undermine the territorial system of ecological stability, they are also obliged to propose measures which contribute to its creation and maintenance. Priority for habitat care should be focused on habitats in protected areas, but in order to maintain the stability and balanced provision of the ES, which will help the landscape's adaptive capacity, attention should also be paid to habitats and ecosystems outside protected areas, with some regulation of their use. In order to promote and preserve biodiversity, it is important to apply the principles of sustainable agriculture and forestry practices in real life, which would also increase the benefits of ecosystems for people in the regions.

Biodiversity is threatened by changes in land use, which poses a significant risk to human society's well-being. The main trend is the increasing intensity of conventional agriculture and forestry, leading to a decline in biodiversity. The decline in traditional farming has resulted in the abandonment and reduction of semi-natural high natural value habitats (Keenleyside and Tucker 2010; Lieskovský et al. 2015). Biodiversity is also threatened by the exploitation and harvesting of natural resources, pollution of the environment and its components, as well as the spread of invasive species.

4.7.5 ES Assessment for the Territory of Slovakia

Research focusing on the assessment of the promotion of species and ecosystem diversity directly as an ecosystem service is quite rare in Slovakia. The current work rather presents the option for assessment of selected ES in case study areas. For territories with the traditional agricultural landscape, the following indicators were used (Špulerová et al. 2014): the importance of habitats (habitat of national or European importance), favourable habitat status and presence of protected and endangered species. The case study of the Trnava functional city area for the assessment of the joint ES biodiversity promotion, life cycle and pest control support used the GreenFrame method, based on expert assessment and synthesis of thematic layers (Mederly et al. 2017).

The proper understanding of this ES was preceded by an assessment of vegetation functions (forest function, non-forest woody or urban vegetation function), while diversity promotion was ranked among natural biotic functions (Brodová 2008; Kontriš 1978; Papánek 1978; Sláviková 1987). Ecological functions of the forest have been investigated in the most detailed way. The key criterion of the basic decision-making system for the assessment of the functional efficiency of forest ecosystems (in the landscape in various ecological-functional and socio-economic conditions) comes in the form of forest structure (nature-based/slightly altered/ greatly altered species, age and space creating optimal trophic conditions for plants). The presence of protected areas, occurrence of rare and endangered species, occurrence of endemic species, seasonal species concentration, degree of environmental degradation and land use were other criteria used (Čaboun et al. 2010).

The capacity of current ecosystems for biodiversity promotion as well as the occurrence of genetically important species can be expressed, in particular, through the following indicators: the presence of significant and rare species (Fig. 4.28), or habitats. The need to preserve the diversity of species and ecosystems is evident

Fig. 4.28 Ecosystems provide space for rare species and their preservation – mountain Apollo (Parnassius apollo). (Author: J. Černecký)

Table 4.7 Input data for capacity, demand and flow assessment of ES biodiversity promotion

Input data/ ES	R7: biodiversity promotion
Capacity	The occurrence of priority and important habitats – map of ecosystems of Slovakia
	Naturalness of habitats (comparison of real forest and non-forest vegetation with potential natural vegetation)
	Significance of the territory in terms of nature and landscape protection – synthesis of territorial nature protection of SR
	Spatial structure of the territory – diversity of the landscape (number of ecosystem types per 1 km^2)
	Habitat status – according to SNC SR data
	Current landscape structure – additional data for territory classification
Demand	Current landscape structure – categorization by demand for this ES (mainly intensively used agricultural areas, forest monocultures)
	State of ecosystems – ecosystems disturbed or in a bad state
	Spatial projection of ecological network – territories with a deficit of significant elements and disturbed ecological stability
Flow	European and nationally significant habitats in a good state – real occurrence
	Locations of occurrence of protected and endangered species, indication species and the like – verified and real occurrence
	Small-scale structures of the agricultural landscape (mosaics) or other important CLS categories
	Ecological network and green infrastructure functional elements

particularly among the experts, who have a greater need to preserve biodiversity for future generations as they are more thoroughly aware of rare and endangered species, as well as their specific requirements and threat factors. Thus, demand can be spatially differentiated, by the real provision of the ES on the basis of ecosystem status, by environmental quality, by the representation of rare and nature-based habitats and so on.

Although it is not realistic to incorporate all the necessary data for the pilot assessment of Slovakia in terms of this ES, the input data are sufficiently representative (Table 4.7). The assessment was based on the *map of ecosystems of Slovakia* (Černecký et al. 2020), created from several available environmental data (especially SNC SR data on habitats and their status, occurrence of protected and endangered species, other data from biotic monitoring, data on forest structure and age, agricultural land use, basic topographic layers). Another input came from the naturalness of vegetation indices, assessed on the basis of comparison of real vegetation and potential natural vegetation. The significance of the territory in terms of nature and landscape protection formed another input – it was expressed on the basis of a combination of different types of protected areas in Slovakia. The biodiversity of the area was assessed as an indicator of the occurrence of the number of different types of ecosystems within a spatial unit of 1 km^2.

The total capacity of the area in terms of promoting species and ecosystem diversity was expressed as a combination of the above-mentioned layers in the relative scale of the landscape's capacity to provide this ES – the assessment result is shown in Fig. 4.29.

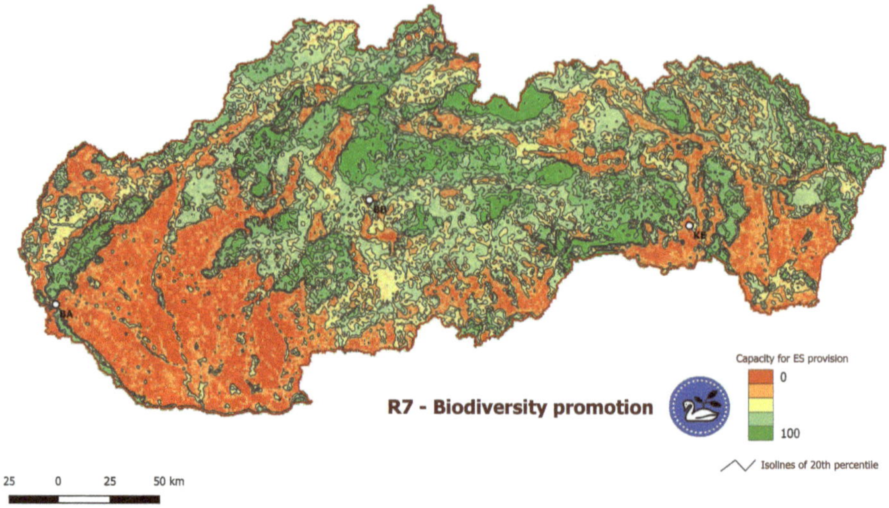

Fig. 4.29 Capacity of the landscape to provide ES biodiversity promotion

The spatial projection of the significance of the territory of Slovakia in terms of biodiversity promotion is logical and obvious – the highest level is typical for part of the mountain and sub-mountain areas, which are mostly forested and belong to the system of protected areas. On the contrary, the lowest level of significance is typical for large agricultural and urbanized areas in the lowlands, partly in the intra-mountain basins of Slovakia. It is these territories which include significant *islands* of biodiversity, which should form the basis for possible further measures to revitalize the landscape.

Table 4.7 also shows the basic indicators which can be used for future expression of the level of ES demand as well as its real use. In this case, the fundamental question is whether biodiversity support is primarily a priority in protected areas or it applies in the entire agricultural and forestry landscape or even in urbanized areas.

From the real production/flow of this ES point of view, it is necessary to focus on the real and verified occurrence and status of important habitats and gene pool sites, on the effect of management and renaturation measures in the landscape or on the functionality of ecological networks in agricultural and urbanized landscapes.

4.8 Life Cycle Maintenance/Pollination (R8)

4.8.1 Definition and Brief Characteristics of ES

The promotion of life cycles and processes includes the promotion of pollen distribution (pollination) as well as the promotion of plant reproduction conditions (seed dispersal), which may include bees, birds, bats, butterflies, flies, flightless animals or wind (Burkhard et al. 2014). Plant pollination is an inevitable and economically important ES which impacts the preservation and promotion of the biodiversity of most wild plants and the fertility, quality and stability of crop production (Kizeková et al. 2016). Based on the Global Pollination Assessment prepared by experts from the Intergovernmental Panel on Biodiversity and Ecosystem Services (IPBES), it is estimated that more than a third of plant production depends on pollinators and approximately 75% of all crops benefit to various extent from pollination by animals, including most vegetables, fruits or spices (Potts et al. 2016). In addition to the effects of pollination on crop production, feed crops for animals or the products of natural ecosystems (e.g. forest fruits), ornamental plant species (e.g. orchids), also benefit or depend on this ES (Schröter et al. 2019).

The abundance and diversity of pollinators in many ecosystems around the world is declining mainly due to the widespread intensification of agriculture, which is mainly linked to excessive use of chemicals, pesticides and monoculture cultivation and the effects of climate change (Benelli et al. 2017). Many studies have shown that the abundance and species diversity of pollinators, as well as pollination intensity, decreases with distance from natural or nature-based habitats (Garibaldi et al. 2011; Ricketts et al. 2008) as they are extensively dependent on habitat options for nesting and flower sources that cannot be found within arable land (Fig. 4.30). The disruption and fragmentation of many natural habitats results mainly from urbanization and increasing intensification of agriculture (Vanbergen et al. 2013).

In response to these negative pressures, various tools and possible approaches are being developed to reverse this state. A diverse array of original pollinators can

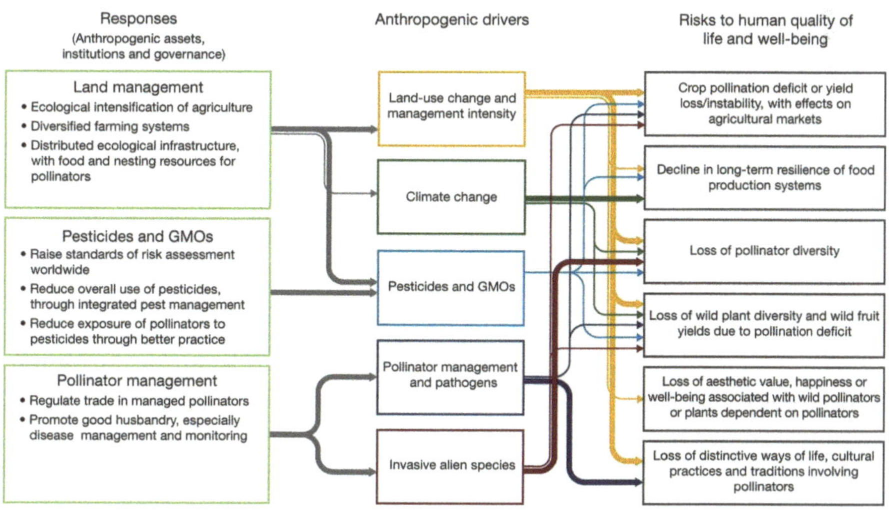

Fig. 4.30 Driving forces, risks and consequences associated with pollinator decline. (Source: Potts et al. 2016)

stabilize population variability between individual years and mitigate the decline in biodiversity of specific pollinator species (Ricketts et al. 2008; Tscharntke et al. 2005). In changing environmental conditions, these species can play an important role in maintaining the ecosystem's resilience (Schröter et al. 2019). Since pollination is representing an ES on which people are dependent through a link to food production, it has often been cited as an example of the economic value of ES (Hanley et al. 2015).

Crop pollination is largely dependent on beekeeping, but bees are often threatened by the combined effect of parasites, diseases and pesticides (Vanbergen et al. 2013). People are sometimes not even very aware of the pollination activity of bees until they see the consequences or evidence directly. People take it rather for granted, and it only becomes apparent when there is a catastrophe associated with the death of bees.

4.8.2 Methods Used to Assess and Identify ES

The first pollination valuations were published already in the 1940s. These assessments preceded the ES concept by nearly 40 years. The publications were not motivated by the protection of pollinators but by the interest in maximizing crop yields. In addition, these valuations focused mainly on honey bees in Europe and North America. Initially, the value was calculated as the total economic value of all crops, the yields of which, albeit minimal, were increased by insect pollinators. This value was the basis for all national assessments up to 1987 and the global assessments up to 2009. In 1987, O'Grady proposed a methodology for the *economic value of insect pollination* to overcome the problems associated with assessment methodology. This methodology includes a crop-specific pollination coefficient, usually based on the crop's biological properties and field research. According to this coefficient, the loss of economic benefits can be calculated if all bees suddenly disappear (Melathopoulos et al. 2015).

Other methods for pollinators assessment include (1) *replacement value* method, where the pollination costs are exchanged for human labour (Muth and Thurman 19951995); (2) *conditional valuation* method based on willingness to pay for the protection of wild pollinators (Mwebaze et al. 2010); or (3) *field services processes* method, which applies to parts of the landscape, its diversity, abundance and yields (Olschewski et al. 2006; Ricketts et al. 2004). All these methods have their limits and limitations, especially in view of the specificities of the territory under investigation and data availability. The environmental component of *pollination* service is represented by the abundance and diversity of pollinators, with the status indicator represented by the number and effectiveness of pollinating species, and the crop dependence on natural pollinators determines the efficiency (UNEP WCMC 2011).

An overview of the research and methods of ES valuation was prepared by Frélichová et al. (2014); economic methods of assessment were mainly used for pollination: benefit transfer and the insect pollination economic value.

Biophysical methods based on mapping or modelling of landscape structure, biodiversity and other natural indicators are also used to assess this ES. The modern modelling tools for ES assessment today include one of the InVEST models focused on the pollinator abundance and crop pollination (Tallis et al. 2011). The primary function of the model is to identify the nesting of wild bees and bumblebees in the landscape based on an embedded raster layer with landscape features and its properties which affect the behaviour of wild bees and bumblebees, supplemented by the list of pollinator species in the landscape and their characteristics, such as the nesting index in the cavities or in the ground, its rate of activity and the range of species to floral resources. The result of the modelling forms a map of the probable occurrence of wild bees with regard to the availability of nesting and flower sources from the surrounding area, which may be helpful, for example, in optimizing agriculture, and can be taken into account in the overall landscape management.

Various indicators are often used to assess this ES:

- In Germany, the proposed ES indicators were divided by supply (share of natural and semi-natural habitats in agricultural land) and demand (representation of pollination-dependant crops) (Albert et al. 2016). In another study, indicators such as the average yield of fruit trees, the density of bees, the proportion of extensively used habitats suitable for pollinator pasture, the distance between crops and pollinators and the area of agricultural crops dependant on pollination were monitored (Rabe et al. 2016).
- The pollination value in Finland was expressed by four indicators: (1) the economic value of pollination based on farmers' incomes for the most economically important crops, such as rapeseed, tomatoes, fruit and berries, as well as wild species; (2) health values by nutrients needed, for example vitamins or phytosterol, which reduces blood cholesterol concentration; (3) the value of the species themselves, which are dependent on the ecological function of pollination; and (4) the social value which affects some popular recreational activities, such as forest fruit harvesting and gardening (Jäppinen and Heliölä 2015).
- Ecosystem potential for ES pollination has been identified in Israel by assessment of two indicators: (1) food resources based on habitat assessment and monitoring (relative abundance of nectar-producing flowers and their flowering period) and (2) nesting possibilities for wild bees based on expert ecosystem assessment (Lotan et al. 2018).
- The national ES assessment in Luxembourg was based on the CICES international classification and the European ES assessment methodology (European Commission 2014a), where capacity indicators (pollination probability expressed by pollinator density) and balance indicators (percentage of pollinated crops expressed as % of area unit) were proposed for this ES (Becerra-Jurado et al. 2016).
- Five key indicators have been proposed for the spatial model to assess ES pollination in Europe (Zulian et al. 2013): (1) suitable nesting sites, (2) availability map of floral resources, (3) spatial range of pollinators, (4) species-specific

parameters in relation to temperature and solar radiation and (5) environmental factors limiting the nesting of pollinators.

- For the ES assessment in Romania, the following indicators were proposed for pollination: (1) structural, area of cultivated rapeseed and productive fruit orchards; (2) functional, abundance of pollinators; (3) evaluating, assessment of pollination deficit expressed by area of crops dependent on pollination; (4) utilitarian, number of beekeepers; and (5) value setting, value of honey produced (NEPA 2017).

4.8.3 The Main Types of Landscape and Ecosystems Which Provide ES

The distribution of the benefits of pollination around the world is very uneven and different in various types of ecosystems, sometimes even within agricultural regions of the same country.

The landscape of Slovakia provides suitable conditions for pollinators and beekeeping. There are widespread forests, which are the original home of bees and form a good-quality bee fodder. In particular, a less influenced landscape with more natural habitats, with the presence of species with a good supply of pollen and nectar, provides suitable conditions for pollinator populations. In particular, forest and scrub habitats are important in terms of pollination quality, as well as orchards. In terms of quantity, these habitats are important: beech and fir-beech forests, lime-oak forests and oak – hornbeam forests. Of the large non-forest areas, these include lowland and submontane hay meadows; other ecosystems of flowering meadows are also important (Fig. 4.31).

Species with a very good supply of pollen and nectar include woody plants such as black locust (*Robinia pseudoacacia*), willows (*Salix purpurea, S. fragilis, S. caprea*), cherries (*Cerasus vulgaris*) even nectarous shrubs like red raspberry (*Rubus idaeus*), currant (*Ribes* sp.), blackthorn (*Prunus spinosa*) and common hazel (*Corylus avellana*). In PG, the following are involved in the high honey-bearing potential: Dutch clover (*Trifolium repens, T. pratense, T. montanum, Medicago lupulina*), dandelion (*Taraxacum officinale*), meadow geranium (*Geranium pratense*), creeping thistle (*Cirsium arvense, C. oleraceum*), common heather (*Calluna vulgaris*), eyebright (*Euphrasia rostkoviana*) and oregano (*Origanum vulgare*). In the case of riparian habitats, these are mainly the stands of white butterbur (*Petasites albus, P. hybridus*), in succession or ruderal communities, for example rosebay willowherb (*Chamerion angustifolium*), honey clover (*Melilotus alba, M. officinalis*) and others.

Fig. 4.31 Species-rich Molinia meadows with flowering Siberian iris (*Iris sibirica*) offer suitable habitats for pollinators. (Author: J. Černecký)

4.8.4 Importance of ES in Terms of Nature and Landscape Protection in Slovakia

Pollination of plants by pollinators contributes to the preservation and promotion of the biodiversity of most wild plants and to the overall functioning of ecosystems (Boyd and Banzhaf 2007; Fisher et al. 2009), as the pollinators are significant for more than 80% of wild plants in temperate climate (Potts et al. 2016). Pollinators find more suitable habitats in the semi-natural habitats, which are part of protected areas with a higher degree of protection against the use and application of chemicals and the like. In keeping the rules which apply to the landscape according to its degree of protection (National Council of the Slovak Republic 2002), nature protection contributes to creating more appropriate conditions for pollinators and preserving the biodiversity of the landscape.

Slovakia is a country with a strong agricultural tradition, while the production of pollinated crops is important for local and regional agriculture. In Slovakia, protection of pollinators is a common agenda of both the Ministry of Environment and Agriculture, who jointly adopt the measures to protect the pollinators. Slovakia is a member of an informal pollination coalition initiated by the Netherlands in December 2016 during the 13th Convention on Biological Diversity in Mexico (MoE SR 2017). The aim of the initiative is to jointly implement national strategies to include new approaches, such as green belts to improve the natural habitat of pollinators; innovations and practices which include promoting bee-friendly farming practices; as well as new partnerships to protect all important pollinators, by supporting diversified farming systems and through the protection, management and

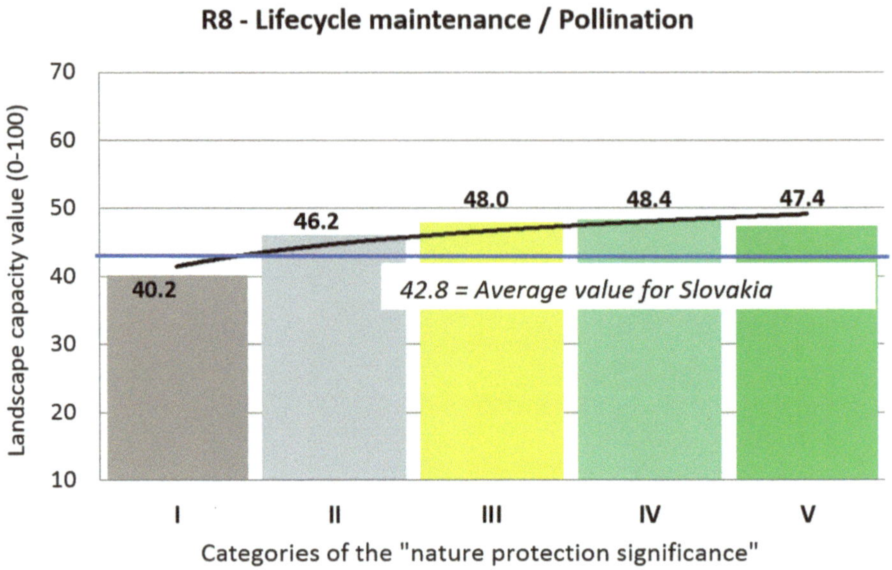

Fig. 4.32 The relationship between ecosystem service R8 and the significance of the territory of Slovakia in terms of nature and landscape protection

restoration of natural habitats in order to increase their extent and connectivity for pollinators.

Specific decisions on how to best ensure ES pollination depend on local circumstances and conditions. In countries where intensive agriculture dominates, measures such as methods of organic agriculture and planting of tree alleys providing floral resources have the largest impact (Schröter et al. 2019).

The fact that the agricultural landscape has its importance in terms of support for pollination can be documented by the *slightly positive* relationship between the significance of the territory in terms of nature and landscape protection and the capacity of Slovakia's landscape to provide this ES (Fig. 4.32). This relationship can be interpreted that there is no clear difference between individual categories of significance of territory in terms of nature and landscape protection – unlike the case of most other supporting and regulatory services, landscape pollination capacity is fairly evenly distributed among all categories.

4.8.5 ES Assessment for the Territory of Slovakia

In Slovakia, the landscape's capacity to provide ES pollination was assessed only in selected model areas using participative methods based on expert estimates for the provisioning capacity of this ES (Bezák and Bezáková 2014; Špulerová et al. 2018). To express plant nectar and pollen reserves within different plant communities,

Fig. 4.33 Gladiolus imbricatus is an attractive wildflower for various types of pollinators. (Author: J. Černecký)

Jurko (1990) suggested calculating the nectar potential, which expresses the percentage of species with pollen and nectar reserves within the overall species composition. The proportion of nectar-producing plants (Fig. 4.33) and nectar reserves within each community is merely an indicative figure, as actual reserves are conditioned by the spatial and physiognomic structure of the plant species and their coverage throughout the community and also by the vegetation phase over a period of time.

The distinction between capacity (supply), demand and the real status of providing pollination is very complex. Potential habitats for pollinators, as well as the abundance and number of pollinator species, can be used to determine the landscape's capacity for pollination support. This can be expressed using a qualitative scale or biophysical units/indicators, such as nesting possibilities density, potential abundance of pollinators and number of bee colonies.

In terms of landscape research and the benefits provided by pollinators for the society, it is also important to examine the environmental factors which affect their

Table 4.8 Input data for capacity, demand and flow assessment of ES life cycle maintenance/pollination

Input data/ES	R8: ES life cycle maintenance/pollination
Capacity	Current landscape structure – favourability of CLS categories for pollinators (reclassification)
	Naturalness of habitats (comparison of real forest and non-forest vegetation with natural potential vegetation) – occurrence of important habitats
	Spatial structure of the territory – diversity of the landscape (number of ecosystem types per 1 km^2) – expressing conditions for the occurrence of pollinators
	Other suitable indicators:
	Data on the use of agricultural land – crops, agricultural land management
	Degree of nature protection, management of protected areas, habitat status
Demand	Current landscape structure – categorization by demand for this ES (especially intensively used agricultural areas and territories with lack of potential and real ES provision)
	Areas of cultivation of special crops and cultures with the need for pollination
	Areas with a deficit of ecologically important elements and disturbed ecological stability – the need to support natural elements
Flow	Small-scale structures of the agricultural landscape (mosaics) or other important categories of agricultural use in terms of honey-bearing potential – real occurrence
	Semi-natural and diverse forest ecosystems, special forest honey-bearing plants – real occurrence
	Occurrence and classification of stress factors – pollution and environmental threat, socio-economic activities – limiting factor of providing this ES

distribution, health and final production. For the pilot assessment of the capacity of the territory of Slovakia, we used the available data, which are sufficiently representative for this ES (Table 4.8). The basic layer was *a map of the current land use* with several categories of agricultural land and forests, which was subsequently reclassified in terms of suitability for pollinators. The assessment was mainly based on the *map of ecosystems of Slovakia* (Černecký et al. 2020), created from several available environmental data (especially SNC SR data on habitats and their status, occurrence of protected and endangered species, other data from biotic monitoring, data on forest structure and age, agricultural land use, basic topographic layers). Another input came from the *naturalness of vegetation*, assessed on the basis of comparison of real vegetation and natural potential vegetation. Spatial structure of the territory in terms of ES promotion was assessed similarly to that of the ES biodiversity promotion with the indicator *biodiversity of the area* based on the occurrence of the number of different types of ecosystems within a spatial unit of 1 km^2.

The overall capacity of the area in terms of supporting life cycles and processes and pollination was expressed as a combination of the above-mentioned layers in a relative scale of the landscape's capacity to provide this ES – the result of the assessment is shown in Fig. 4.34. Unlike most other regulatory and supporting services, the spatial interpretation of individual landscape capacity categories in this ES is not so obvious. Although the highest values are achieved in larger forest and mountain

Fig. 4.34 Capacity of the landscape to provide ES life cycle maintenance/Pollination

complexes, the overall picture is like a mosaic, and perhaps with the exception of the larger areas of the southern Slovak lowlands, all categories of landscape capacities are evenly balanced. This shows the importance of the agricultural land for provisioning of this ES.

Demand for pollination can be expressed by the necessary amount of pollinators and pollinated area (achieving the *desired* value, taking into account the area, population and ecosystem status). Demand for pollination services is the result of farmers' decisions to grow crops that are dependent on pollination (Lautenbach et al. 2011), or the amount and spatial distribution of crops, garden and wild plants requiring pollination (Burkhard et al. 2014).

In terms of comparison of demand and real provision of this ES, it is necessary to say that the greatest demand is typical for lowland areas with dominant agricultural production, with many crops being dependant on pollination. The most favourable situation in terms of demand and production is mainly in Northern and Central Slovakia with a high proportion of forest and permanent grassland habitats, where the production of this ES is clearly exceeded by demand. In particular, agricultural and forest habitats are among the most important consumers of this ES, and therefore the agricultural sector is largely dependent on it. This is particularly evident in the region of Western Slovakia, with the demand being higher than the production of this ES from a national perspective. This deficit is mainly offset by beekeepers with their colonies. In regions where demand exceeds the production of this ES, there also exists a need to increase the presence of semi-natural ecosystems which provide suitable habitats for pollinators and also there is a need for creation of the suitable conditions to support beekeepers and eliminate factors that cause mortality or decrease of numbers of beehives.

4.9 Pest and Disease Control (R9)

4.9.1 Definition and Brief Characteristics of ES

Pest and disease control expresses the *ability of ecosystems to regulate pests and diseases through genetic variations of plants and animals*, thereby contributing to improving the ability of ecosystem resistance and mitigating the risk of spreading of diseases/pests and invasive/non-native species (Burkhard et al. 2014). The structure of the landscape influences local diversity and ecosystem processes, including mutual interactions of species and habitats, characterized by the different dynamics of these communities. The species can be associated with certain communities but can also move between different communities, both natural and anthropogenic (Tscharntke et al. 2005).

While the plant biodiversity is involved through energy and nutrient flows in regulatory functions of natural ecosystems, this form of control is gradually disappearing from the landscape as a result of agricultural intensification associated with environmental pollution, biodiversity loss, synantropization, habitat degradation and the creation of artificial ecosystems that are unstable and requiring constant human intervention and cause increased economic burden (Rusch et al. 2016). In addition to the intensification of agriculture, the prevalence and spread of diseases and pests is also influenced by human population growth, accidental introduction of pests and pathogens, land management and the impact of farming on wildlife. Intensively used agroecosystems are deprived of the natural regulatory capacity to support their own soil fertility and pest control, so costly external inputs need to be supplied to crops. These interventions can reduce the quality of life due to reduced soil, water and food quality if these are contaminated with pesticides and/or nitrates. Commercial preparation of seedbed and mechanized planting has replaced natural seed dispersal methods. Chemical pesticides have replaced the natural processes of control of weed, insects and pathogen populations; genetic manipulation replaces the natural processes of plant evolution and selection. At present and in the future, changes in climatic and hydrological conditions will increasingly affect the spread of diseases and pests. Changes in ecosystems can directly affect the number of human pathogens and can alter a number of disease transmitters (e.g. mosquitoes), as well as affect the incidence of pests and diseases of crop and cattle. In terms of landscape management, increasing the exchange of species between agroecosystems and semi-natural ecosystems can have both positive and undesirable interactions (Fig. 4.35).

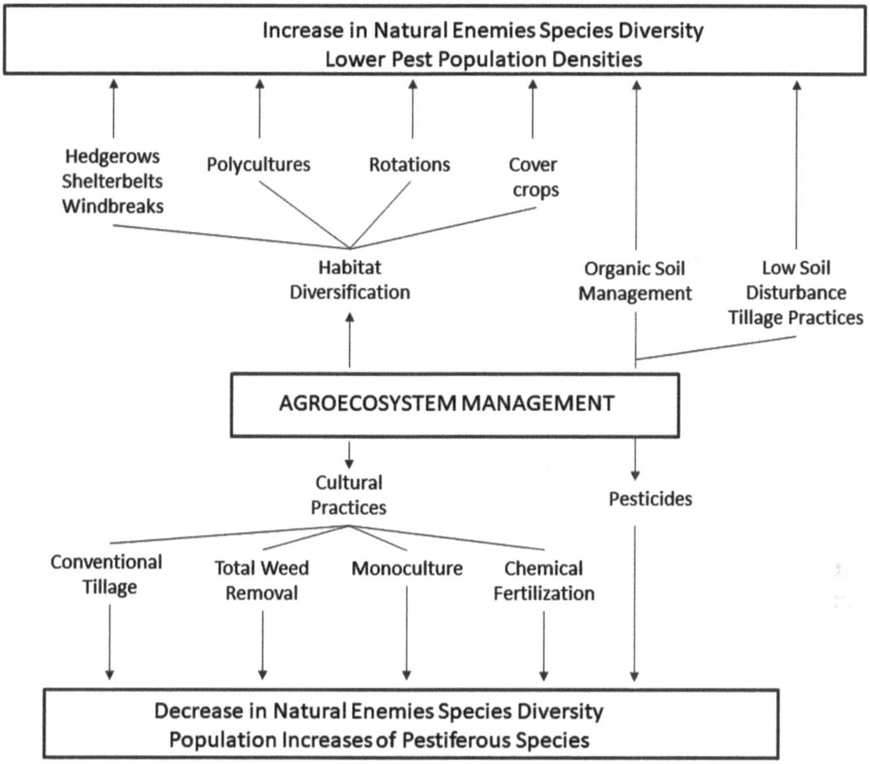

Fig. 4.35 Impact of agroecosystem management and related cultural practices on the biodiversity of natural enemies and the number of insect pests. (Source: Altieri 1999)

Similarly, for forest ecosystems, areas of planted monocultures are characterized by reduced stability and ability of these ecosystems for restoration, as manifested by, for example, in calamities. One of the most extensive ones in recent years was the windstorm in 2004, which affected the territories of the Vysoké and Nízke Tatry, Horehronie, Orava, Kysuce and Spiš (Kunca et al. 2014).

Original habitats and species can be also negatively affected by *non-native invasive plant or animal species*, which do not have their original area of distribution in Slovakia and have the potential to spread rapidly. In the case of their mass distribution, they significantly change the habitat character, threaten the native plant species and create homogeneous monocenoses. Some of them, such as the giant hogweed (*Heracleum mantegazzianum*) or common ragweed (*Ambrosia artemisiifolia*), are the causes of human health problems such as allergies and skin diseases.

4.9.2 Methods Used to Assess and Identify ES

As with other regulatory services, it is sometimes difficult to distinguish between the potential and the actual flows/contribution of ecosystems to the provision of a given ES. Therefore, service flow indicators have been proposed for some ES, including pest and disease control, to prevent the emergence and spread of pests and diseases, in proportion to capacity. Of course, the magnitude and effects of prevented events are difficult to measure in most cases, and the identification of definite location for the demand for a particular ES may be problematic (Burkhard et al. 2014).

Many studies provide examples of the assessment of this ES based on biophysical indicators, for example:

- Status indicator (service rate): number and effectiveness of pest control species
- Performance indicator (service sustainability): reduction of crop pests, human and animal diseases (UNEP WCMC 2011)
- Forest interactions with other habitats (list of functions, species); effects caused by forest change (benefits and loss of functions) – practical assessment of forest quality at the landscape level (Dudley et al. 2012)
- Density of small-scale structures on agricultural land or in special crops – national ES assessment in Germany (Grunewald et al. 2016)

Frélichová et al. (2014) have prepared an overview of research and methods of ES valuation, and for this ES, the following methods are used: biophysical assessment prepared in the form of a review (summary of data/indicators using biophysical metrics) or economic assessment methods (benefit transfer, contingent valuation). The benefit transfer method represents the application of values in monetary terms, with the values obtained by research for specific studies and applied to another, similar study. The contingent valuation method is used to determine the value of an ecosystem by identifying how much respondents are willing to pay for certain ecosystem benefits or services.

In another study (Farber et al. 2006), two methods have been proposed for the economic assessment of the ES: avoided cost – and production approach. When using the avoided cost method, the value derived from research is the cost of preventing or reducing environmental risk. The production approach assessment is based on the values of indirect benefits which could be caused by pests and diseases on agricultural production.

As is the case with other ES, the *GreenFrame method* was also used for this ES, based on a wide range of spatial data set (analytical maps) grouped into themes in combination with expert assessment (Kopperoinen et al. 2014).

4.9.3 The Main Types of Landscape and Ecosystems Which Provide ES

Considering the potential for the provision of this service, the natural and semi-natural habitats in the neighbourhood of agroecosystems or other anthropogenic areas are particularly important. Several studies of interspecies relations show that the diversity and abundance of beneficial species of herbivorous insects and predators, and thus the regulatory function of ecosystems, are higher, for example in ecotone communities, extensive orchards, natural grassland and mosaic-cultivated fields than in an intensively farmed large-scale agricultural landscape (Altieri 2004; Andow 1991; Collins et al. 1996; Offenberg 2015). These habitats located at the frontier of arable land plots can contribute to the control of pests and diseases of farm animals and plants, to the reduction of disease transmitters, human pathogens and the like.

The greatest benefit of this ES is visible in areas where supply and demand are in an approximate balance, i.e., for example, in a *diversified agricultural or urbanized landscape* with sufficient ecosystem representation which offer habitats for many animal species, thus creating a potential for promoting natural pest control (Schröter et al. 2019). With an increasing number of enemies, it is believed that biological pest control is also increasing.

It is therefore obvious that the spatial distribution of areas with a higher capacity for provision of this ES will be very closely correlated with the occurrence of the areas suitable for the provision of ES biodiversity promotion. Nature and nature-based ecosystems with proper status have the highest ability to participate in pest and disease control, and their functionality decreases with the disturbed state.

4.9.4 Importance of ES in Terms of Nature and Landscape Protection in Slovakia

Natural and semi-natural habitats in landscape used by humans perform a balancing function by creating the conditions and space for nesting of relevant bird species, the space for the protection of small animals and the conditions for activity of pollinators, thus contributing to mitigating the risk of spreading diseases/pests and invasive/non-native species. They attenuate the negative effects of anthropogenic activity in the landscape and its components, thus contributing, in particular, to increasing the stability of the landscape and improving the ability of ecosystem restoration. Habitats of national or European importance are often small-scale protected areas in the midst of an intensively used agricultural landscape, thus largely fulfilling the function of pest and disease control. These protected areas and their protection zones (declared/non-declared) are subject to a higher level of territorial protection, which sets the conditions for the practical protection of the landscape and eliminates negative activities, affecting the habitat status, such as the

Fig. 4.36 Traditional agricultural mosaics with a diversified land structure and the presence of non-forest woody vegetation on the plot boundaries significantly contribute to pest and disease control (Hlboké nad Váhom). (Author: J. Špulerová)

application of chemical and fertilization. In this way, it ensures the conditions for better fulfilment of this ES in the landscape.

The effectiveness of this ES can be enhanced by *supporting the management of landscape diversity*, preservation and establishment of elements of ecological interest (Fig. 4.36), by creating bio-belts (increasing the stability of the landscape and the ecosystems themselves). This can also be helped by highlighting the importance of biodiversity (predators, antagonist parasites, soil microflora and microfauna) in providing crop protection and soil fertility *and developing agroecological technologies and systems*, which emphasize the conservation/regeneration of biodiversity, soil, water and other resources. Such measures are urgently needed to meet the growing range of socio-economic and environmental challenges and to enhance the ecosystems provide habitats for pest and disease control. Thus, it can be stated that the relationship of this ecological function with the principles of nature and landscape protection is complementary and mutually supportive (see Fig. 4.37).

For the prospective restoration of ecosystems with the aim to improve the quality of this ES, it would be necessary to *improve the condition of forest ecosystems*, as a significant part of forests is threatened by calamities due to deteriorated health. Similarly, it is appropriate to promote an increase in the presence of semi-natural habitats within the agricultural landscape, with these habitats then serving as refuges, and to eliminate any danger to these habitats from the spreading of non-native

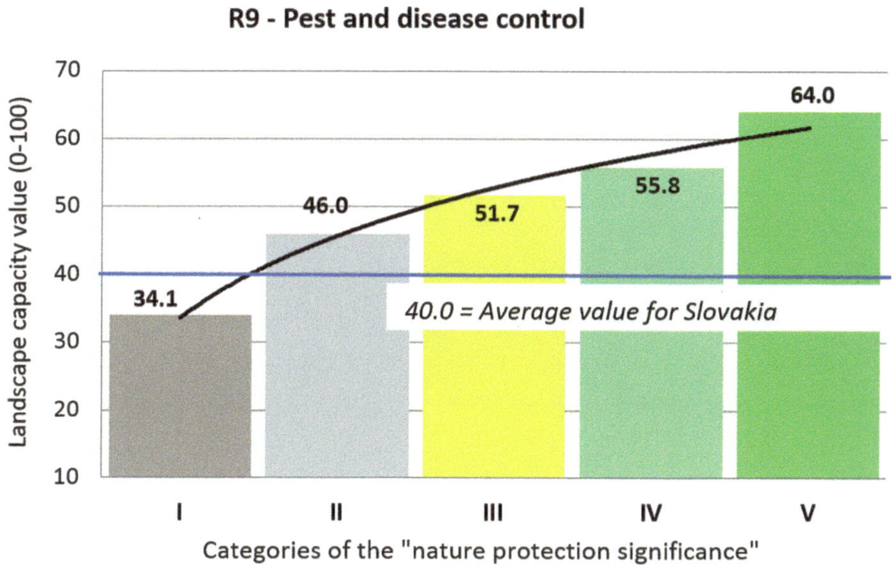

Fig. 4.37 The relationship between ecosystem service R9 and the significance of the territory of Slovakia in terms of nature and landscape protection

invasive species, as well as to promote a territorial geo-diversity (including abiotic environment) and diversity of land use.

4.9.5 ES Assessment for the Territory of Slovakia

So far, the practical assessment of this ES in Slovakia is rare. The joint ES – biodiversity, life cycles and pest control promotion – was assessed in a case study of the Trnava City functional area using the *GreenFrame* method, with the use of expert assessment and qualitative assessment of multiple map layers (Mederly et al. 2017).

Similar to the ES R7 assessment, the determinant factor here is the type of ecosystems and their status, as well as the selected positive and negative factors of the environment (Fig. 4.38). The pilot assessment of the capacity of the territory of Slovakia in terms of ES pest and disease control followed the ES R7 biodiversity promotion – as the data for the landscape's real state of health are not available, input indicators were selected from this ES. However, it should be noted that these are closely related ecosystem functions and services, the principles of which have a common basis in a favourable ecosystem state.

The main input into the assessment included the layers of naturalness of vegetation and habitat status in terms of quality and management. The spatial landscape structure was assessed by the diversity of the landscape based on the number of different types of ecosystems within the spatial unit of 1 km^2 (Table 4.9). The total capacity of the territory for the regulation of pests and diseases was expressed as a combination of these layers – the result of the assessment is shown in Fig. 4.39.

Fig. 4.38 A healthy and resilient ecosystem (in good state) can eliminate pests (including spruce bark beetle) by itself and prevent property damage. (Author: J. Černecký)

Table 4.9 Input data for capacity, demand and flow assessment of ES pest and disease control

Input data/ ES	R9: pest and disease control
Capacity	The naturalness of habitats (comparison of real forest and non-forest vegetation with natural potential vegetation)
	Spatial landscape structure – diversity of the landscape (number of ecosystem types per 1 km^2)
	Habitat status – according to SNC SR data
	Current landscape structure – additional data for territory classification
Demand	Current landscape structure – categorization by demand for this ES (populated areas, ruderal areas, intensively used agricultural areas, forest monocultures)
	State of ecosystems – ecosystems disturbed or in a bad state
	Environmental quality – damaged or disturbed areas (air quality, environmental hygiene, etc.)
	The occurrence of invasive species, allergens and the like
	Population distribution – densely populated areas, areas with increased occurrence of allergies, etc.
Flow	Habitat classification (forest and non-forest) – significant habitats, the occurrence of indication species and the like
	Small-scale structures of agricultural land (mosaics) or other CLS categories
	Areas with a real implementation of agri-environmental measures

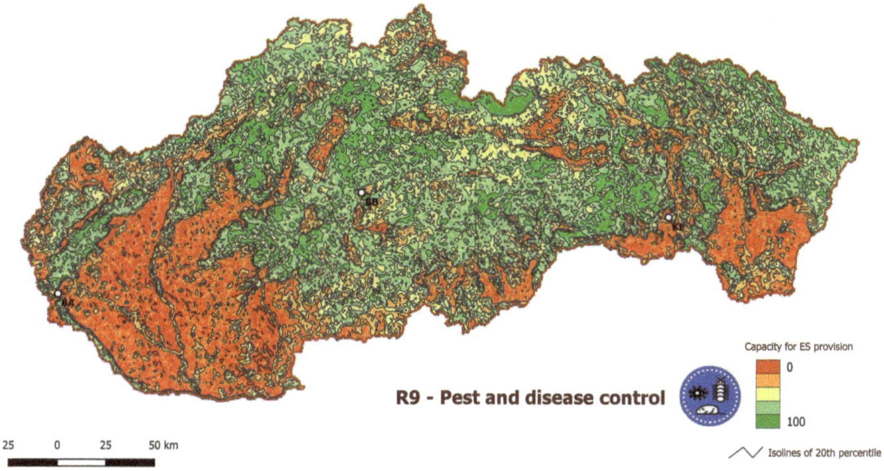

Fig. 4.39 Capacity of the landscape to provide ES pest and disease control

The spatial projection of the significance of Slovakia's territory from the point of view of this ES (Fig. 4.39) is very similar to that of ES biodiversity promotion, although the resulting image is more mosaic. The highest capacity is typical for mountains and sub-mountain areas, while the lowest capacity is documented in large agricultural and urban areas of Slovakia. The basis for the reconstruction of the biodiversity and regulatory functions of the landscape in these areas should come from the already mentioned *islands* of biodiversity and higher ecological quality, which are mainly bound to hydric and forest ecosystems.

The largest demand for this ES is obvious in settlements with mainly anthropogenic ecosystems and in areas of lowlands and basins with intensive agricultural activity, which are characterized by the low share of non-forest habitats and by being unstable. An important consumer of this ES is the agricultural land itself and thus the agricultural sector, which requires additional energy and constant human intervention to ensure the landscape stability and for prevention of the spread of pests and diseases. In regions where demand for this ES exceeds the production, it is necessary to increase the proportion of ecosystems which produce this service (e.g. a greater share of forest ecosystems) and to increase the functional biodiversity of agroecosystems (through the creation of multifunctional field margins – bio-belts on the arable land). Such practices are mainly applied in areas of organic farming which are supported by the Rural Development Program as part of the pillar II.

In the first pillar of the Common Agricultural Policy in relation to the conservation of biodiversity, the promotion of direct *greening* payments, linked to the implementation of the following procedures, contributes to such conservation: diversification of crops, permanent grassland maintenance and ecological focus areas (EFA). In ecological focus areas larger than 15 ha, a minimum of 5% of the area needs to be set apart for the following elements: fallow land; terraces; landscape features such as a solitary trees, row trees, small woods and hedges; buffer zones; areas with fast-growing species, with intermediate crops or green cover; or areas with nitrogen-binding crops. However, it should be noted that such a share is still insufficient in terms of maintaining ecosystems and hence the quality of the provision of this ES.

4.10 Maintenance of Soil Formation and Composition (R10)

4.10.1 Definition and Brief Characteristics of ES

Soil is the top layer of the weathered bedrock of the earth's crust containing water, air and living organisms. It is divided into horizontal layers with specific physical, chemical and biological properties. Individual layers have different ecological functions and functions related to human activities (Article 1 of the Principles of State Soil Policy of the SR, approved in 2001). The soil belongs to the essential component of the landscape necessary for life development and thus ecosystems. The above-mentioned document declares that soil is a common "wealth" of the citizens of the state and the heritage for future generations. It is an essential and non-renewable natural resource and forms an integral part of Earth's ecosystems. It is and will remain the basis of Slovakia's environmental, ecological, economic and social potential and must, therefore, be carefully protected from damage and unjustified reduction in its area and volume.

Soils represent complex ecosystems which consist of living and inanimate matter with lots of interconnections between them. The diversity and abundance of life in the soil is greater than in any other ecosystem. A small volume of soil can contain billions of different organisms that play a crucial role in soil quality to support plant growth. In addition to its participation in various biogeochemical cycles and nutrient exchange, the soil provides many other important ES (Schröter et al. 2019).

Soil formation is a long-term process of weathering of the bedrock and accumulation of organic particles. The soil environment is part of the main nutrient cycle in the environment – these being essential for life processes of organisms (e.g. N, S, P, C). Nutrients are decomposed and recycled in this process, changing forms, becoming available to plants and animals and for the ecosystem cycle. Biological fertility of soil is an important attribute of total soil fertility. The beneficial effects of soil organisms on the fertility of agricultural land are clear and obvious (available online: www.agroporadenstvo.sk). Soil processes such as the nutrient cycle, water cycle and biological activity promote soil formation and thus contribute to the development of soil properties and the provision of soil natural capital reserves. ES maintenance of soil formation is also dependent on the bedrock, climate, vegetation, time and territory in which they are located (Dominati et al. 2010).

Fertile and healthy soil is necessary for *ecosystem functioning and for food production*. Also, undisturbed soils can store and retain large amounts of carbon, which in turn has a beneficial effect on climate regulation. Soils are essential because they perform a number of essential functions in the landscape, such as nutrient cycle, water regulation, habitat protection and biodiversity, filtering and mitigating, as well as the stability of the area itself. The presence of dead biomass (necromass) is considered a good indicator of the ability of soil to perform these basic functions. In this context, the function of necrophages, invertebrates and organisms ensuring decomposition of organic material is very important. They are actively involved in interactions which develop in the soil between physical, chemical and biological processes. A comprehensive analysis of invertebrate activities shows that they can be seen as the best indicators of soil quality and at the same time should be considered as a resource to be managed to improve the provision of ES by agroecosystems (Lavelle et al. 2006). For example, Pavlík et al. (2015) experimentally followed the decomposition of various size wood fractions with saprophytic fungi – such fungi can be used for quicker decomposition of waste/unused dendromass and thus a faster intake of nutrients to forest land. Neher et al. (2012) explain the rate of decomposition of woody material by macrofauna (e.g. arthropods).

Due to the processes associated with pedogenesis, carbon is deposited in the topsoil layer, and the overall physical properties of the soil are improved. Significant benefits can be achieved with the proper functioning of processes associated with pedogenesis and the maintenance of soil quality – for example, this includes the need to reduce exogenous agricultural inputs (Becerra-Jurado et al. 2016).

The most important and most valued soil function is the provision of the substrate for plant growth. Almost all food production and a substantial portion of the raw materials and energy recovered is provided by plants growing on the soil. The importance of soil is still understood and assessed today especially in the context of agroecosystems, which provide for agricultural production of crops. However, as mentioned above, its quality is also equally important for the growth of other plants and woody biomass, as well as for several regulatory and supporting services (storage and distribution of carbon and other chemicals, regulation of runoff conditions and erosion processes, filtration and water purification, ensuring conditions for soil biodiversity, etc.). Soil properties are therefore very important not only for the functioning of the agricultural landscape ecosystems but also for other types of ecosystems which provide other ecosystem functions and related services for humans.

Soil quality regulation is a primary ecosystem function which plays a key role in providing regulatory services through storage and decomposition of organic substances, mediating the exchange of gases into the atmosphere, storing, decomposing and transforming materials, such as nutrients and contaminants, and regulating water flows. These supporting functions are largely related to the role of ecosystems in soil quality regulation and contribute significantly to other final ecosystem services, such as climate regulation, detoxification and purification as well as crop production and other products (e.g. fibres), growth of trees and others vegetation and peat formation (UK NEA 2011).

Simply put, in the context of ecosystem functions and services, we understand ES maintenance of soil formation and composition as the creation and maintenance of favourable conditions for the long-term provision of non-production soil functions.

4.10.2 Methods Used to Assess and Identify ES

ES Maintenance of soil formation and composition, in its essence, is a *strictly* scientific domain, so the biophysical methods are dominant in its assessment.

According to Pérez-Soba, Harrison et al. (2015) and Czúcz et al. (2018), the most important indicators for this ES include:

- Physical properties of soil: carbon stock, water capacity, soil structure
- Biological parameters: organic matter content, soil nutrients, biological recovery, above-ground biomass
- Process of pedogenetic processes: mineralization, decomposition, nutrient cycle
- Character of soil-forming bedrock

Other suitable indicators include, for example, land use management (agricultural production, forestry, urbanization activities), environmental pollution (soil, surface and groundwater contamination) and the share of organic farming (Maes et al. 2014).

Most of the national or regional assessments of the soil-related ES focus on its production characteristics and are predominantly assessed in terms of agroecosystems. According to Schröter et al. (2019), one of the assessment approaches includes the integration of the current understanding of soil-related processes into appropriate models to describe the dynamics of soil functions and related indicators. These models are usually designed for specific soil-related processes, such as carbon dynamics in soil, water flow in soil, soil compaction or greenhouse gas emissions. Change of soil functions corresponds to change of these properties, which in turn are influenced by land management practices. Another approach to soil assessment is to characterize the soil as a specific combination of its functional properties. What is traditionally known as *soil type* can be translated into a combination of functional properties (e.g. bulk density, organic carbon content, functional soil biota diversity).

The maintenance of soil formation and composition was investigated in more detail in the national assessment of, for example, Finland (Jäppinen and Heliölä 2015) and Great Britain (UK NEA 2011). In both cases, biophysical proxy indicators were used, namely, the functional diversity of soil organisms, nutrient cycle (Finland) or soil carbon, soil chemistry and heavy metal soil pollution (UK).

To a lesser extent, economic (monetary) methods are also used for the assessment of this ES (Frélichová et al. 2014) – it is possible to financially quantify the value of carbon or nutrients stored in the soil. Sandhu et al. (2010), on the other hand, assesses the soil on the basis of the market value of the earthworm-aerated topsoil layer and the mineralization estimates based on the market value of nitrogen that would otherwise have to be supplied externally to the soil. It is a *price-substitution* method which quantifies selected ecosystem functions and economically reflects the situation, where these functions would have to be artificially replaced. Colombo et al. (2006), in turn, followed the *willingness to pay for the ES* method to estimate the average price for a specific erosion regulation project, depending on its quality. Bond et al. (2011) estimated the value of soil from the cost of irrigation, which prevents erosion and loss of nitrogen.

Non-production functions and a system of qualitative and financial assessment of agricultural soils in Slovakia were investigated by Vilček (2014). In particular, he used biophysical assessment methods based on a number of indicators expressing the capacity of the soil to accumulate water (field water capacity), immobilize risk elements (sorption potential and content of risk elements), immobilize organic pollutants (C content and humus quality, clay content, soil depth, precipitation) and transform organic pollutants (C content and humus quality, clay content, air temperature). He also prepared a so-called soil environmental potential index (SEPI), as well as the financial expression of the main soil environmental functions.

For the ES soil formation itself, analytical indicators are the most important of this system – the content and quality of the organic soil component, clay content, soil depth, water capacity and soil sorption potential.

4.10.3 The Main Types of Landscape and Ecosystems Which Provide ES

For the proper functioning of the soil ecosystem, a healthy environment is necessary, without the presence of any serious negative factors (pollution and damage to the environment, intensive land use influencing natural processes and soil regime). That is why *nature and nature-based ecosystems* provide a suitable environment for the creation and circulation of nutrients and support for the main ecosystem functions associated with the soil environment (Fig. 4.40). These ecosystems include, in particular, *forest areas and grassland ecosystems* of large size, where space and time are provided for these processes to stay uninterrupted.

The reservoir of nutrients and their transformation media for the transfer to the soils is represented by *watercourses, water bodies and wetlands* – in this respect, they are very important for natural ecosystems with good ecological status. On the other hand, intensively used ecosystems (especially agroecosystems but also commercially used forest stands) are typically affecting the natural regime and the flow of nutrients – the use of natural resources, the disruption of natural cycles and the input of additional energy. The natural biological activity of the soil has been replaced by industrial fertilizers, chemicals and mechanization which have changed the chemical and physical properties of the soil, its biological activity and the like. The totally altered soil environment and the related disrupted main soil functions are present in urbanized ecosystems, where anthropogenic processes dominate.

Fig. 4.40 Decomposers – invertebrates involved in the decomposition of dead plant biomass, soil formation and sanitation activities for ecosystem cleaning. (Author: J. Černecký)

Approximately 12 mil. ha of soil is threatened annually in the world due to its desertification and degradation (Schröter et al. 2019).

On the other hand, proper management practices, especially in agroecosystems, can support the soil's biological fertility and gradually improve its physical-chemical properties. Soil environmental functions are increasingly taking on an economic as well as ethical and moral dimension. Assessment and valuation of the soil's capacity to perform vital tasks can significantly help in its necessary protection, especially in the case of thoughtless land take or anthropic interventions in the landscape (Vilček 2014).

4.10.4 Importance of ES in Terms of Nature and Landscape Protection in Slovakia

As specified in the assessment of ES *biomass for food production*, modern agriculture (and in part also the forestry) has become a threat to the proper functioning of ecosystems and thus to the fulfilment of non-production ecosystem functions and soil-related services. Therefore, intensive agriculture is perceived as a negative factor in relation to nature and landscape protection.

On the other hand, the promotion of non-production soil functions is generally consistent with the main objectives of nature and landscape protection. A well-functioning soil environment provides a number of non-production functions which directly or indirectly support the ecosystem's natural protection function. Similarly, it is clear that active nature protection, including the promotion of ecological and

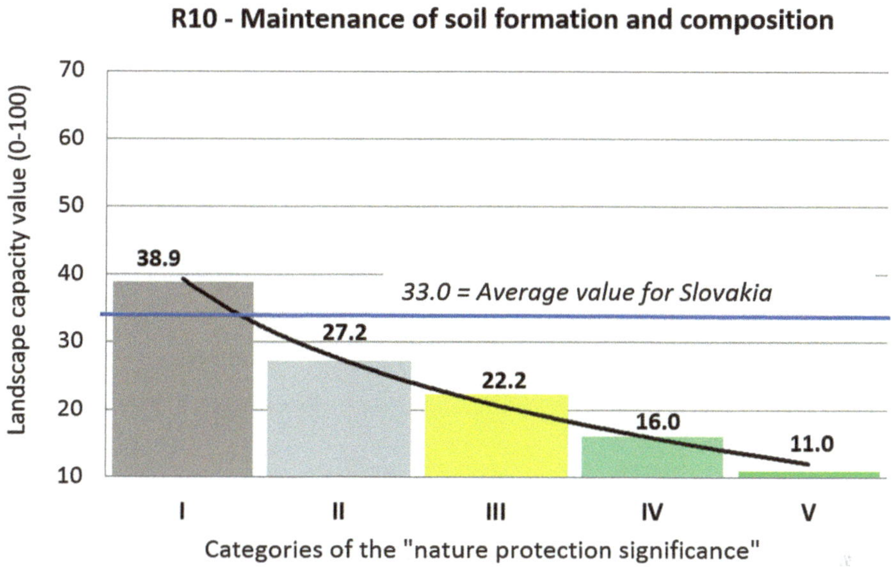

Fig. 4.41 The relationship between ecosystem service R10 and the significance of the territory of Slovakia in terms of nature and landscape protection

non-production functions of the landscape, is also in line with ES maintenance of soil formation and composition. Nevertheless, the current relationship between the landscape's capacity for provision of this ES and significance of the territory in terms of nature and landscape protection is characterized by a negative correlation (see Fig. 4.41) – this is mainly the result of the low capacity of mountain soils to fulfil this function due to their unfavourable physical and partly chemical conditions.

Management of soils and the promotion of their non-production functions should be an important priority for the agricultural sector, at the same level as the production function of agricultural and forestry landscape. This fact has been largely reflected in European sectoral policies – a system is in place to support the non-production functions and services of the rural landscape (Fig. 4.42) in the form of subsidy schemes.

The current *Rural Development Program 2014–2020* also focuses on sustainable management, restoration, conservation and enhancement of ecosystems, promotion of resource efficiency and support for the transition to a low-carbon climate-resilient economy. Direct support from the resources of the European Agricultural Fund for Rural Development (EAFRD) can be used, for example, for organic farming or for agri-environment-climate measures. The resources of the state budget can be used as direct support, for example, for climate and environment-friendly farming practices (available online: www.apa.sk). The purpose of these subsidies is thus indirectly to support several regulatory ES (anti-erosion, water management, soil protection) and supporting functions of the agricultural landscape (in particular, biodiversity promotion, pest and disease protection, support of soil fertility), to

Fig. 4.42 Agroforestry landscape in the Stará Turá region. (Author: D. Štefunková)

some extent support for cultural ES (agrotourism and recreation, preserving historical structures in the agricultural landscape).

4.10.5 ES Assessment for the Territory of Slovakia

The capacity of Slovakia's agricultural soils to perform environmental functions was expressed by Vilček (2014) with the so-called soil environmental potential index (for more details please see previous text). The total index was created as a combination of four sub-indices, which take the value from 1 (very high capacity) to 5 (very low capacity), and is also expressed in a point scale of 20–100 points. The average point value reflecting the capacity of Slovakia's agricultural land to provide environmental functions is 55.3 points. This highest average point value was achieved by agricultural land in the Nitra Region (72 points) and Šaľa District (82 points). The lowest average point value was recorded in the Prešov Region (48 points) and in the district of Košice 1 (42 points) and Gelnica (41 points). This index can serve as a spatial indicator expressing the heterogeneity of the capacity of Slovakia's agricultural land to provide selected environmental functions.

As this assessment is only valid for agricultural land and is more focused on hygiene functions (and there is no other assessment available for the territory of

Table 4.10 Input data for capacity, demand and flow assessment of ES maintenance of soil formation and composition

Input data/ ES	R10: maintenance of soil formation and composition
Capacity	Soil production potential (agricultural and forest soils)
	Soil filtration capacity
	Climatic conditions (especially the moisture balance)
	Relief – slope inclination
Demand	Intensively used agricultural (partly also forested) areas with depletion of nutrients and carbon
	Degraded and contaminated areas with infertile or hygienically harmful soils
	Disturbed ecosystems or ecosystems in a bad state
Flow	Territories with favourable soil characteristics (based on pedological surveys and analyses)
	Undisturbed areas with balanced use of soil resources (agroforestry areas)
	Territories with a practical implementation of agri-environmental measures and with the improvement of soil properties in a natural way

Slovakia), it is necessary to look for other indicators to express the capacity (potential) of the landscape to fulfil the supporting function of improving the formation and natural composition of the soil.

The pilot assessment of the capacity of the territory of Slovakia in terms of this supporting function followed the assessment of ES P1 biomass production – agricultural crops, where the productivity (fertility) of the soil was assessed in particular. This assessment was also taken as a basis for this ES, supplemented by filtering and buffering capacity of soils and correction coefficients expressing properties of relief (slope inclination) and climate (moisture balance). It is a simplified assessment of the total capacity of the area in terms of supporting pedogenesis and soil fertility (Table 4.10). The result was expressed as a combination of the above-mentioned layers in a relative scale – spatial differentiation is shown in Fig. 4.43.

From the resulting map, it is evident that lowland areas with favourable soil properties (depth, nutrient content, flat relief, suitable climate) have the largest capacity, while the lowest capacity is achieved by mountain areas with low capacity to support pedogenesis and related processes. However, sub-mountain and transitional areas with average landscape capacities are also important – with lightly disturbed environment and lower anthropogenic pressures (Fig. 4.42) than in the case of intensively exploited areas, which have a relatively good preconditions for a significant fulfilment of this ecosystem function.

Demand for ES maintenance of soil formation and composition is determined by the intensity of use and the state of the environment – the greatest demand is present in areas with the largest pressure to use the soil's production function (intensively used agricultural and forestry areas) or in areas with disturbed environment (degraded and contaminated areas, disturbed ecosystems).

The real use (flow) of this ES is, in turn, given either by the natural processes improving or promoting important soil characteristics and fertility or by appropriate

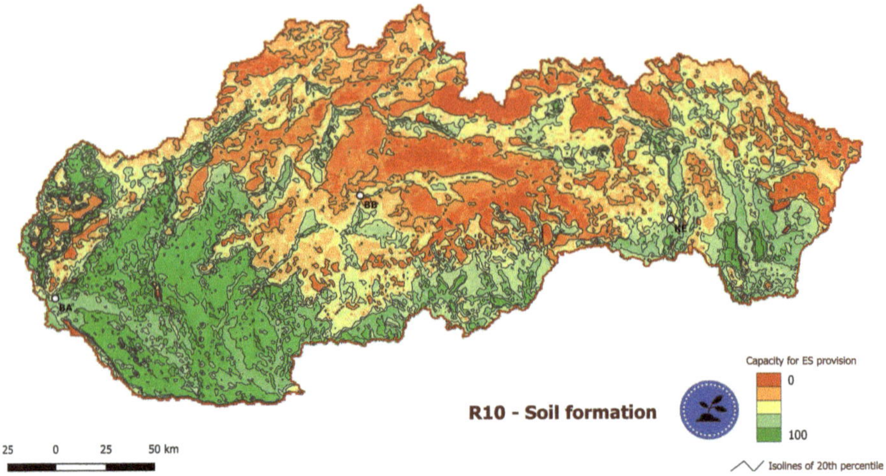

Fig. 4.43 Capacity of the landscape to provide ES maintenance of soil formation and composition

use and management of (in particular) the agricultural landscape. However, such indicators are likely to be very difficult to obtain at the national level.

References

Albert, C., Bonn, A., Burkhard, B., Daube, S., Dietrich, K., Engels, B., Frommeri, J., Goetzl, M., Gret-Regamey, A., Job-Hoben, B., Koellner, T., Marzelli, S., Moning, C., Mueller, F., Rabe, S.-E., Ring, I., SChwaiger, E., Schweppe-kraft, B., & Wuestemann, H. (2016). Towards a national set of ecosystem service indicators: Insights from Germany. *Ecological Indicators, 61*, 38–48. https://doi.org/10.1016/j.ecolind.2015.08.050.

Ali-toudert, F., & Mayer, H. (2007). Effects of asymmetry, galleries, overhanging façades and vegetation on thermal comfort in urban street canyons. *Solar Energy, 81*, 742–754. https://doi.org/10.1016/j.solener.2006.10.007.

Altieri, M. A. (1999). The ecological role of biodiversity in agroecosystems. *Agriculture, Ecosystems & Environment, 74*(1–3), 19–31. https://doi.org/10.1016/S0167-8809(99)00028-6.

Altieri, M. A. (2004). Linking ecologists and traditional farmers in the search for sustainable agriculture. *Frontiers in Ecology and the Environment, 2*(1), 35–42. https://doi.org/10.1890/1540-9295(2004)002[0035:leatfi]2.0.co;2.

Andow, D. (1991). Yield loss to arthropods in vegetationally diverse agroecosystems. *Environmental Entomology, 20*(5), 1228–1235. https://doi.org/10.1093/ee/20.5.1228.

Antal, J. (2005). *Protierózna ochrana pôdy*. Slovakia: Slovak University of Agriculture in Nitra.

Barančok, P., & Barančoková, M. (2015). Distribution of the traditional agricultural landscape types reflecting geological substrate and slope processes in the Kysuce Region. *Ekologia Bratislava, 34*(4), 339–355. https://doi.org/10.1515/eko-2015-0031.

Barka, I., & Rybár, R. (2003). Identification of snow avalanche trigger areas using GIS. *Ecology, 22*(2), 182–194.

Bebi, P., Kulakowski, D., & Rixen, C. (2009). Snow avalanche disturbances in forest ecosystems—State of research and implications for management. *Forest Ecology Management, 257*, 1883–1892. https://doi.org/10.1016/j.foreco.2009.01.050.

Becerra-Jurado, G., Philipsen, C., & Kleeschulte, S. (2016). *Mapping and assessing ecosystems and their services in Luxembourg – Assessment results*. Luxembourg: Le Gouvernment du Grand -Duché de Luxemburg. https://doi.org/10.13140/rg.2.1.4924.5841.
Benelli, G., Benvenuti, S., Scaramozzino, P. L., & Canale, A. (2017). Food for honeybees? Pollinators and seed set of Anthyllis barba-jovis L. (Fabaceae) in arid coastal areas of the Mediterranean basin. *Saudi Journal of Biological Sciences, 24*(5), 1056–1060. https://doi.org/10.1016/j.sjbs.2017.01.018.
Berry, R., Livesley, S. J., & Ayeb, L. (2013). Tree canopy shade impacts on solar irradiance received by building walls and their surface temperature. *Building and Environment, 69*, 91–100. https://doi.org/10.1016/j.buildenv.2013.07.009.
Bezák, P., & Bezáková, M. (2014). Landscape capacity for ecosystem services provision based on expert knowledge and public perception (case study from the North West Slovakia). *Ekologia Bratislava, 33*(4), 344–353. https://doi.org/10.2478/eko-2014-0031.
BISE. (2019). *The Biodiversity Information System for Europe*. https://biodiversity.europa.eu/topics/ecosystem-services. Accessed 4 Apr 2019.
Bond, C. A., Hoag, D. L., & Kipperberg, G. (2011). Agricultural producers and the environment: A stated preference analysis of Colorado corn producers. *Canadian Journal of Agricultural Economics, 59*(1), 127–144.
Boyd, J., & Banzhaf, S. (2007). What are ecosystem services? The need for standardized environmental accounting units. *Ecological Economics, 63*(2–3), 616–626. https://doi.org/10.1016/j.ecolecon.2007.01.002.
Brauman, K. A., Daily, G. C., Duarte, T. K., & Mooney, H. A. (2007). The nature and value of ecosystem services: An overview highlighting hydrologic services. *The Annual Review of Environment and Resources, 32*, 6.1–6.32.
Brodová, M. (2008). *Ekonomické hodnotenie mimoprodukčných funkcií poľnohospodárstva v podmienkach*. Dissertation, Slovak Academy of Sciences.
Bruijnzeel, L. A. (2004). Hydrological functions of tropical forests: Not seeing the soil for the trees? *Agriculture, Ecosystems & Environment, 104*, 185–228. https://doi.org/10.1016/j.agee.2004.01.015.
Bujnovský, R. (2018). Estimation of benefits from the actual use of inland water ecosystem services in the Slovak Republic. *Ekológia, 37*(3), 201–218. https://doi.org/10.2478/eko-2018-0017.
Burkhard, B., & Maes, J. (Eds.). (2017). *Mapping ecosystem services. Bulgaria*. Sofia: Pensoft Publishers.
Burkhard, B., Kroll, F., Nedkov, S., & Müller, F. (2012). Mapping ecosystem service supply, demand and budgets. *Ecological Indicators, 21*, 17–29. https://doi.org/10.1016/j.ecolind.2011.06.019.
Burkhard, B., Kandziora, M., Hou, Y., & Müller, F. (2014). Ecosystem service potentials, flows and demands – Concepts for spatial localisation, indication and quantification. *Landscape Online, 34*(1), 1–32. https://doi.org/10.3097/LO.201434.
Čaboun, V., Tutka, J., Moravčík, M., Kovalčík, M., Sarvašová, Z., Schwarz, M., & Zemko, M. (2010). *Uplatňovanie funkcií lesa v krajine*. Zvolen: National Forest Centre.
Capotorti, G., Ortí, M. M. A., Anzellotti, I., Azzella, M. M., Copiz, R., Mollo, B., & Zavattero, L. (2015). The MAES process in Italy: Contribution of vegetation science to implementation of European Biodiversity Strategy to 2020. *Plant Biosystems – An International Journal Dealing with all Aspects of Plant Biology, 149*(6), 949–953. https://doi.org/10.1080/11263504.2015.1095253.
Černecký, J., Gajdoš, P., Špulerová, J., Halada, Ľ., Mederly, P., Ulrych, L., Ďuricová, V., Švajda, J., Černecká, Ľ., Andráš, P., & Rybanič, R. (2020). Ecosystems in Slovakia. *Journal of Maps, 16*, 28–35. https://doi.org/10.1080/17445647.2019.1689858.
Černý, J. (2012). *Modelování povodňové vlny při destrukci vybrané vodní nádrže*. Olomouc: Palacký University Olomouc.
Coates, D., Pert, P. L., Barron, J., Muthuri, C., Nguyen-Khoa, S., Boelee, E., & Jarvis, D. I. (2013). Water-related ecosystem services and food provision. In E. Boelee (Ed.), *Managing water and agroecosystems for food provision* (pp. 29–41). Wallingford: CABI. http://www.iwmi.cgiar.org/

Publications/CABI_Publications/CA_CABI_Series/Managing_Water_and_Agroecosystems/chapter_3-water-related_ecosystem_services_and_food_security.pdf. Accessed 2 Mar 2019.

Collins, K. L., Wilcox, A., Chaney, K., & Boatman, N. D. (1996). Relationships between polyphagous predator density and overwintering habitat within arable field margins and beetle banks. In: *British crop protection conference: Pests and diseases*, Brighton, UK, 18–21 November 1996.

Colombo, S., Calatrava-Requena, J., & Hanley, N. (2006). Analysing the social benefits of soil conservation measures using stated preference methods. *Ecological Economics, 58*, 850–861. https://doi.org/10.1016/j.ecolecon.2005.09.010.

Czúcz, B., Arany, I., Potschin-Young, M., Bereczki, K., Kertész, M., Kiss, M., Aszalós, R., & Haines-Young, R. (2018). Where concepts meet the real world: A systematic review of ecosystem service indicators and their classification using CICES. *Ecosystem Services, 29*, 145–157. https://doi.org/10.1016/j.ecoser.2017.11.018.

De Groot, R. S. (1992). *Functions of nature: Evaluation of nature in environmental planning, management and decision making*. Groningen: Wolters Noordhoff.

De Groot, R. S., Wilson, M. A., & Boumans, R. M. (2002). A typology for the classification, description and valuation of ecosystem functions, goods and services. *Ecological Economics, 41*(3), 393–408. https://doi.org/10.1016/s0921-8009(02)00089-7.

Dominati, E., Patterson, M., & Mackay, A. (2010). A framework for classifying and quantifying the natural capital and ecosystem services of soils. *Ecological Economics, 69*(9), 1858–1868. https://doi.org/10.1016/j.ecolecon.2010.05.002.

Dudley, N., & Stolton, S. (2003). *Running pure: The importance of forest protected areas to drinking water*. Washington, DC/Gland: World Bank/WWF Alliance for Forest Conservation and Sustainable Use.

Dudley, N., Schlaepfer, R., Jackson, W., Jeanrenaud, J.-P., Stolton, S., Schlaepfer, R., Jackson, W., Jeanrenaud, J.-P., & Stolton, S. (2012). *Forest quality: Assessing forests at a landscape scale*. London: Routledge.

ENVIROPORTÁL. (2018). *Indikátory stavu a ochrany biodiverzity*. https://www.enviroportal.sk/indicator/301?langversion=sk. Accessed 2 Mar 2019.

European Commission. (2012). *A blueprint to safeguard Europe's water resources*. https://eur-lex.europa.eu/legal-content/SK/TXT/PDF/?uri=CELEX:52012DC0673&from=EN. Accessed 2 Mar 2019.

European Commission. (2014a). *System of environmental-economic accounting 2012 experimental ecosystem accounting*. New York: European Commission Organisation for Economic Co-operation.

European Commission. (2014b). *Mapping and assessment of ecosystems and their services: Indicators for ecosystem assessments under Action 5 of the EU Biodiversity Strategy to 2020*. http://ec.europa.eu/environment/nature/knowledge/ecosystem_assessment/pdf/2ndMAESWorkingPaper.pdf. Accessed 2 Mar 2019.

Farber, S., Costanza, R., Childers, D. L., Erickson, J., Gross, K., Grove, M., Hopkinson, C. S., Kahn, J., Pincetl, S., Troy, A., Warren, P., & Wilson, M. (2006). Linking ecology and economics for ecosystem management. *BioScience, 56*(2), 121–133. https://doi.org/10.1641/0006-3568(2006)056[0121:LEAEFE]2.0.CO;2.

Fernando, S. M., Montes, C., López, M. B., González, A. J., Aguado, M., Benayas, J., Piñeiro, C., Navacerrada, J., Zorrilla, P., Llorente, M. G., Iniesta, I., Oteros, E., Palomo, I., Santiago, C. S., Alcorlo, P., Vidal, R., & Suárez, L. (2013). *Ecosystems and biodiversity for human wellbeing. Synthesis of the key findings*. Madrid: Biodiversity Foundation of the Spanish Ministry of Agriculture, Food and Environment.

Fischer, A., Fischer, H. S., & Lehnert, U. (2012). Avalanche creating high structural and floristic diversity in mountain mixed forests in the Alps. *Biodiversity and Conservation, 21*, 643–654. https://doi.org/10.1007/s10531-011-0204-z.

Fisher, B., Turner, R. K., & Morling, P. (2009). Defining and classifying ecosystem services for decision making. *Ecological Economics, 68*(3), 643–653. https://doi.org/10.1016/j.ecolecon.2008.09.014.

Fowler, D., Pilegaard, K., Sutton, M. A., Ambus, P., et al. (2009). Atmospheric composition change: Ecosystems-atmosphere interactions. *Atmospheric Environment, 43*, 5193–5267. https://doi.org/10.1016/j.atmosenv.2009.07.068.

Frélichová, J., Vačkář, D., Pártl, A., Loučková, B., Harmáčková, Z. V., & Lorencová, E. (2014). Integrated assessment of ecosystem services in the Czech Republic. *Ecosystem Services, 8*, 110–117. https://doi.org/10.1016/j.ecoser.2014.03.001.

Gallay, I. (2010). Využitie modelovania povrchového odtoku pri hodnotení zraniteľnosti krajiny vo vzťahu k vybraným prírodným hrozbám. *Geografický časopis, 62*(2), 109–125.

Garibaldi, L. A., Steffan-Dewenter, I., Kremen, C., Morales, J. M., Bommarco, R., Cunningham, S. A., Carvalheiro, L. G., Chacoff, N. P., Dudenhoeffer, J. H., Greenleaf, S. S., Holzschuh, A., Isaacs, R., Krewenka, K., Mandelik, Y., Mayfield, M. M., Morandin, L. A., Potts, S. G., Ricketts, T. H., Szentgyoergyi, H., Viana, B. F., Westphal, C., Winfree, R., & Klein, A. M. (2011). Stability of pollination services decreases with isolation from natural areas despite honey bee visits. *Ecology Letters, 14*(10), 1062–1072. https://doi.org/10.1111/j.1461-0248.2011.01669.x.

Georgi, J. N., & Dimitriou, D. (2010). The contribution of urban green spaces to the improvement of environment in cities: Case study of Chania, Greece. *Building and Environment, 45*(6), 1401–1414. https://doi.org/10.1016/j.buildenv.2009.12.003.

Giarratano, M. C., et al. (2018). *Comitato Capitale Naturale*. Secondo Rapporto sullo Stato del Capitale Naturale in Italia. Roma. https://www.minambiente.it/comunicati/il-secondo-rapporto-sullo-stato-del-capitale-naturale-italia. Accessed 6 Apr 2019.

Granec, M., Šurina, B., et al. (1999). *Atlas pôd SR*. Bratislava: Soil Science and Conservation research Institute.

Grešová, L. (2010). *Modelovanie veternej erózie v k.ú. Šaľa a Kráľová n. Váhom*. Dissertation, Slovak Academy of Sciences.

Grizzetti, B., Lanzanova, D., Liquete, C., & Reynaud, A. (2015). *Cook-book for water ecosystem service assessment and valuation* (JRC Science and Policy Report). Luxembourg: European Commission. https://ec.europa.eu/jrc/en/science-update/cook-book-water-ecosystem-service-assessment-and-valuation. Accessed 10 Apr 2019.

Grizzetti, B., Lanzanova, D., Liquete, C., Reynaud, A., & Cardoso, A. C. (2016). Assessing water ecosystem services for water resource management. *Environmental Science & Policy, 61*, 194–203. https://doi.org/10.1016/j.envsci.2016.04.008.

Grunewald, K., Herold, H., Marzelli, S., Meinel, G., Richter, B., Syrbe, R.-U., & Walz, U. (2016). Concept of national indicators for ecosystem services in Germany further development, types of classes and sheet of indicators. *Naturschutz und Landschaftsplanung, 48*, 141–152. https://doi.org/10.3897/oneeco.2.e14021.

Haberl, H., Erb, K. H., Krausmann, F., Gaube, V., Bondeau, A., Plutzar, C., Gingrich, S., Lucht, W., & Fischer-Kowalski, M. (2007). Quantifying and mapping the human appropriation of net primary production in earth's terrestrial ecosystems. *Proceedings of the National Academy of Sciences, 104*(31), 12942–12947. https://doi.org/10.1073/pnas.0704243104.

Haines-Young, R., & Potschin, M. (2013). *CICES V4.3-revised report prepared following consultation on CICES Version 4*, August – December 2012. Nottingham: Centre for Environmental Management, University of Nottingham. https://cices.eu/content/uploads/sites/8/2018/01/Guidance-V51-01012018.pdf. Accessed 6.

Halada, Ľ., Halabuk, A., & Gajdoš, P. (2011). Poľnohospodárska krajina s vysokými prírodnými hodnotami. *Životné prostredie, 45*(1), 12–16.

Hanley, N., Breeze, T. D., Ellis, C., & Goulson, D. (2015). Measuring the economic value of pollination services: Principles, evidence and knowledge gaps. *Ecosystem Services, 14*, 124–132. https://doi.org/10.1016/j.ecoser.2014.09.013.

Hreško, J. (1998). Avalanche hazard of the Tatra high mountain landscape. *Folia Geographica, 2*, 348–352.

Hreško, J., Bugár, G. (1999). Súčasný vývoj lavínových morfosystémov Belianskych Tatier. Paper presented at Prínos a perspektívy tatranského národného parku v ochrane prírodného dedičstva Karpát, Štátne lesy TANAPu, Stará Lesná, 25–28 Nov. 1998.

Hunter, B. A., Livesley, S., & Williams, N. S. G. (2012). *Responding to the urban heat island: A review of the potential of green infrastructure*. Victorian Centre for Climate Change Adaptation Research. http://www.vcccar.org.au/sites/default/files/publications/VCCCAR%20Urban%20Heat%20Island%20-WEB.pdf. Accessed 8 Apr 2019.

Janák, M., Černecký, J., & Saxa, A. (Eds.). (2015). *Monitoring of animal species of community interest in the Slovak Republic. Results and assessment in the period of 2013–2015*. Banská Bystrica: State Nature Conservancy of the Slovak Republic.

Jäppinen, J. P., Heliölä, J. (Eds.). (2015). Towards a sustainable and genuinely green economy. The value and social significance of ecosystem Services in Finland (TEEB for Finland). In *Synthesis and roadmap. Helsinki: The Finnish Ministry of Environment*. Helsinki.

Jeníček, M. (2009). *Modelování pruběhu extrémních povodní v kontextu krajinních změn a integrované protipovodňové ochrany*. Prague: Charles University in Prague, Faculty of Science.

Jurko, A. (1990). *Ekologické a socio-ekonomické hodnotenie vegetácie*. Bratislava: Píroda.

Kadlec, V. (2010). *Využitie zrážkovo-odtokových modelov v bystrinných povodiach*. Disertation, Technical University of Zvolen.

Kaletová, T., & Šinka, K. (2012). Simulácia odtoku vody z extrémneho dažďa pomocou prostredia GIS a CN-metódy. *Acta hydrologica Slovaca, 13*(2), 324–333.

Keeler, B. L., Polasky, S., Brauman, K. A., Johnson, K. A., Finlay, J. C., O'Neill, A., Kovacs, K., & Dalzell, B. (2012). Linking water quality and well-being for improved assessment and valuation of ecosystem services. *Proceedings of the National Academy of Sciences, 109*(45), 18619–18624. https://doi.org/10.1073/pnas.1215991109.

Keenleyside, C., & Tucker, G. (2010). *Farmland abandonment in the EU: An assessment of trends and prospects: Report prepared for WWF*. London: Institute for European Environmental Policy.

Kizeková, M., Čunderlík, J., Dugátová, Z., Makovníková, J., Kanianska, R., Jaďuďová, J., Jančová, Ľ., & Pálka, B. (2016). *Agroekosystémové služby a súčasný stav trávnych porastov v Slovenskej republike*. Banská Bystrica: NPPC VÚTPHP Banská Bystrica, NPPC SSCRI Bratislava, UMB Banská Bystrica.

Kleeschulte, S., & Ruf, K. (2016). *Mapping and assessing ecosystems and their services in Luxembourg – Amendment to final synthesis report*. Luxembourg.

Kobza, J., Barančíková, G., Makovníková, J., Styk, J., Širáň, M., & Vojtáš, J. (2005). *Návrh regulačných pôdoochranných opatrení z výsledkov monitoringu pôd SR*. Bratislava: Soil Science and Conservation Research Institute.

Konarska, J., Lindberg, F., Larsson, A., Thorsson, S., & Holmer, B. (2013). Transmissivity of solar radiation through crowns of single urban trees—Application for outdoor thermal comfort modelling. *Theoretical and Applied Climatology, 117*, 363–376. https://doi.org/10.1007/s00704-013-1000-3.

Konôpka, J. (2012). Manažment hydrických funkcií lesov. *Lesnícky časopis – Forestry Journal, 58*(2), 129–135.

Kontriš, J. (1978). Fyziognomicko-ekologická typizácia lesov a krovín a ich funkcia v krajine. *Problémy ekológie krajiny, 23*, 81–122.

Kopperoinen, L., Itkonen, P., & Niemelä, J. (2014). Using expert knowledge in combining green infrastructure and ecosystem services in land use planning: An insight into a new place-based methodology. *Landscape Ecology, 29*(8), 1361–1375. https://doi.org/10.1007/s10980-014-0014-2.

Kumar, P. (Ed.). (2010). *The economics of ecosystems and biodiversity: Ecological and economic foundations*. London: UNEP/Earthprint.

Kunca, A., Galko, J., & Zúbrik, M. (2014). Aké významné kalamity v posledných rokoch postihli naše lesy? *Letokruhy*. http://www.lesmedium.sk/casopis-letokruhy/2014/letokruhy-2014-07/ake-vyznamne-kalamity-v-poslednych-rokoch-postihli-nase-lesy. Accessed 4 Apr 2019.

Lal, R. (2004). Soil carbon sequestration impacts on global climate change and food provision. *Science, 304*(5677), 1623–1627. https://doi.org/10.1126/science.1097396.

Lautenbach, S., Kugel, C., Lausch, A., & Seppelt, R. (2011). Analysis of historic changes in regional ecosystem service provisioning using land use data. *Ecological Indicators, 11*(2), 676–687. https://doi.org/10.1016/j.ecolind.2010.09.007.

Lavelle, P., Tedaens, T., Aubert, M., Barot, S., Blouin, M., Bureau, F., Mergerie, P., Mora, P., & Rossi, J.-P. (2006). Siol invertebrates and ecosystem services. *European Jounal of Siol Biology, 42*(1), S3–S15. https://doi.org/10.1016/j.ejsobi.2006.10.002.

Lavorel, S., Grigulis, K., Lamarque, P., Colace, M.-P., Garden, D., Girel, J., Pellet, G., & Douzet, R. (2011). Using plant functional traits to understand the landscape distribution of multiple ecosystem services. *Journal of Ecology, 99*(1), 135. https://doi.org/10.1111/j.1365-2745.2010.01753.x.

Lee, S. H., & Park, S. U. (2008). A vegetated urban canopy model for meteorological and environmental modelling. *Bound-Lay Meteorol, 126*(1), 73–102. https://doi.org/10.1007/s10546-007-9221-6.

Lee, H., Holst, J., & Mayer, H. (2013). Modification of human-biometeorologically significant radiant flux densities by shading as local method to mitigate heat stress in summer within urban street canyons. *Advances in Meteorology, 2013*, 1–13. https://doi.org/10.1155/2013/312572.

Lieskovský, J., Bezák, P., Špulerová, J., Lieskovský, T., Koleda, P., Dobrovodská, M., Buergi, M., & Gimmi, U. (2015). The abandonment of traditional agricultural landscape in Slovakia – Analysis of extent and driving forces. *Journal of Rural Studies, 37*, 75–84. https://doi.org/10.1016/j.jrurstud.2014.12.007.

Loh, J., & Harmon, D. (2005). A global index of biocultural diversity. *Ecological Indicators, 5*(3), 231–241. https://doi.org/10.1016/j.ecolind.2005.02.005.

Lotan, A., Kost, R., Mandelik, Y., Peled, Y., Chakuki, D., Shamir, S. Z., & Ram, Y. (2018). National scale mapping of ecosystem services in Israel – Genetic resources, pollination and cultural services. *One Ecosystem, 3*, e25494. https://doi.org/10.3897/oneeco.3.e25494.

Luyssaert, S., Inglima, I., & Jung, M. (2007). CO_2 balance of boreal, temperate, and tropical forests derived from a global database. *Global Change Biology, 13*(12), 2509–2537. https://doi.org/10.1111/j.1365-2486.2007.01439.x.

Maes, J., Paracchini, M. L., Zulian, G., Dunbar, M. B., & Alkemade, R. (2012). Synergies and trade-offs between ecosystem service supply, biodiversity, and habitat conservation status in Europe. *Biological Conservation, 155*, 1–12. https://doi.org/10.1016/j.biocon.2012.06.016.

Maes, J., Teller, A., Erhard, M., Liquete, C., Braat, L., Berry, P., Egoh, B., Puydarrieux, P., Fiorina, C., Santos, F., Paracchini. M.L., Keune, H., Wittmer, H., Hauck, J., Fiala, I., Verburg, P.H., Condé, S., Schägner, J.P., San Miguel, J., EstreguiL, C., Ostermann, O., Barredo, J.I., Pereira, H.M., Stott, A., Laporte, V., Meiner, A., Olah, B., Royo Gelabert, E., Spyropoulou, R., Petersen, J.E., Maguire, C., Zal, N., AchilleoS, E., Rubin, A., Ledoux, L., Brown, C., Raes, C., Jacobs, S., Vandewalle, M., Connor, D., Bidoglio, G. (2013). Mapping and Assessment of Ecosystems and their Services. An analytical framework for ecosystem assessments under action 5 of the EU biodiversity strategy to 2020. Luxembourg: Publications office of the European Union.

Maes, J., Teller, A., Erhard, M., Murphy, P., Paracchini, M. L., Barredo, J. I., Grizzetti, B., Cardoso, A., Somma, F., Petersen, J.-E., Meine, R. A., Gelabert, E. R., Zal, N., Kristensen, P., Bastrup-Birk, A., Biala, K., Romao, C., Piroddi, C. H., Egoh, B., Fiorina, C. H., Santo, F., Naruševičius, V., Verboven, J., Pereira, H., Bengtsson, J., Gocheva, K., Marta-Pedroso, C., Snäll, T., Estreguil, C. H., San Miguel, J., Braat, L., Grêt-Regamey, A., Perez-Soba, M., Degeorges, P., Beaufaron, G., Lillebø, A., Abdul Malak, D., Liquete, C., Condé, S., Moen, J., Östergård, H., Czúcz, B., Drakou, E. G., Zulian, G., & Lavalle, C. (2014). *Mapping and Assessment of Ecosystems and their Services (MAES)*. Indicators for ecosystem assessments under Action 5 of the EU Biodiversity Strategy to 2020 2nd Report – Final. European Union.

Maes, J., Fabrega, N., Zulian, G., Barbosa, A., Vizcaino, P., Ivits, E., Polce, C. H., Vandecasteele, I., Rivero, I. M., Guerra, C., Castillo, P. C., Vallecillo, S., Baranzelli, C., Barranco, R., Silva, B. F., CrisonI, C. H. J., & Trombetti, M. (2015). *Mapping and assessment of ecosystems and their services: Trends in ecosystems and ecosystem services in the European Union between 2000 and 2010*. Luxembourg: Publications Office of the European Union.

Maes, J., Teller, A., Erhard, M., Grizzetti, B., Barredo, J. I., Paracchini, M. L., Condé, S., Somma, F., Orgiazzi, A., Jones, A., Zulian, A., Petersen, J. E., Marquardt, D., Kovacevic, V., Abdul Malak, D., Marin, A. I., Czúcz, B., Mauri, A., Loffler, P., Bastrup-Birk, A., Biala, K., Christiansen, T., & Werner, B. (2018). *Mapping and assessment of ecosystems and their services: An analytical framework for ecosystem condition*. Luxembourg: Publications office of the European Union.

Markov, B., & Nedkov, S. (2016). Mapping of erosion regulation ecosyste services. In Bandrova T., & Konečny, M. (Eds.), *Proceedings 6th international conference on cartography and GIS*, Albena, Bulgaria, 13–17 June 2011.

Matero, J., & Saastamoinen, O. (2007). In search of marginal environmental valuations – Ecosystem services in Finnish forest accounting. *Ecological Economics, 61*(1), 101–114. https://doi.org/10.1016/j.ecolecon.2006.02.006.

MEA – Millennium Ecosystem Assessment. (2005). *Ecosystems and human well-being: Wetlands and water synthesis*. Washington, DC: World Resources Institute.

Mederly, P., Bezák, P., Lieskovský, J., Halabuk, A., Izakovičová, Z., & Dobrucká, A. (2017). Vybrané metódy hodnotenia ekosystémových služieb – projekt OpenNESS a prípadová štúdia Trnava. *Životné Prostredie, 51*, 205–212.

Melathopoulos, A. P., Cutler, G. C., & Tyedmers, P. (2015). Where is the value in valuing pollination ecosystem services to agriculture? *Ecological Economics, 109*, 59–70. https://doi.org/10.1016/j.ecolecon.2014.11.007.

Minár, J., & Tremboš, P. (1994). Prírodné hazardy – hrozby, niektoré postupy ich hodnotenia. Acta Facultatis rerum naturalium Universitatis Comenianae. *Geographica, 35*, 173–194.

MoE SR. (2017). *Envirorezort sa pripojil k iniciatíve za záchranu opeľovačov – Ministerstvo životného prostredia Slovenskej republiky*. http://www.minzp.sk/tlacovy-servis/tlacove-spravy/tlacove-spravy-2017/tlacove-spravy-marec-2017/envirorezort-pripojil-k-iniciative-za-zachranu-opelovacov.html. Accessed 6 Apr 2019.

Mooney, H., Larigauderie, A., Cesario, M., Elmqvist, T., Hoegh-Guldberg, O., Lavorel, S., Mace, G. M., Palmer, M., Scholes, R., & Yahara, T. (2009). Biodiversity, climate change, and ecosystem services. *Current Opinion in Environmental Sustainability, 1*, 46–54. https://doi.org/10.1016/j.cosust.2009.07.006.

MPSR NLC. (2018). *Správa o lesnom hospodárstve v Slovenskej republike za rok 2017 – Zelená správa*. http://www.mpsr.sk/index.php?navID=123&id=13656. Accessed 4 Apr 2019.

Muth, M. K., & Thurman, W. N. (1995). Why support the price of honey? *Choices, 10*(2), 19–22. https://doi.org/10.22004/ag.econ.131245.

Mwebaze, P., Marris, G. C., Budge, G. E., Brown, M., Potts, S. G., Breeze, T. D., & Macleod, A. 2010. *Quantifying the value of ecosystem services: A case study of honeybee pollination in the UK*. Paper presented at 12th annual BIOECON conference "from the wealth of nations to the wealth of nature: Rethinking economic growth", Venice, Centro Culturale Don Orione Artigianelli, 27–28 September 2010.

Nakaohkubo, K., & Hoyano, A. (2011). *Development of passive design tool using 3D-Cad compatible thermal simulation – Prediction of indoor radiation environment considering solar shaving by surrounding trees and buildings*. Paper presented at proceedings of building simulation 2011: 12th conference of International Building Performance Simulation Association, Sydney, 14–16 November 2011.

National Council of the Slovak Republic – NR SR (2002). *Zákon NR SR 543/2002 Z. z. o ochrane prírody a krajiny v znení neskorších predpisov*.

Neher, D. A., Weicht, T. R., & Barbercheck, M. E. (2012). Linking invertebrate communities to decomposition rate and nitrogen availability in pine forest soils. *Applied Soil Ecology, 54*, 14–23. https://doi.org/10.1016/j.apsoil.2011.12.001.

NEPA. (2017). *Assessment of ecosystems and ecosystem services in Romania*. Romania: NEPA, NINA, ROSA, WWF Romania.

Offenberg, J. (2015). REVIEW: Ants as tools in sustainable agriculture. *Journal of Applied Ecology, 52*(5). https://doi.org/10.1111/1365-2664.12496.

Olschewski, R., Tscharntke, T., Benítez, P., Schwarze, S., & Klein, A.-M. (2006). Economic evaluation of pollination services comparing coffee landscapes in Ecuador and Indonesia. *Ecology and Society, 11*(1), 7. https://doi.org/10.5751/ES-01629-110107.

Palm, C. H., Blanco-Canqui, H., Declerckc, F., Gatere, L., & Grace, P. (2014). (2014). Conservation agriculture and ecosystem services: An overview. *Agriculture. Ecosystems and Environment, 187*, 87–105. https://doi.org/10.1016/j.agee.2013.10.010.

Papánek, F. (1978). *Teória a prax funkčne integrovaného lesného hospodárstva*. Bratislava: Príroda.

Parker, N., Naumann, E.-K., Medcalf, K., Haines-Young, R., Potschin, M., Kretsch, C., Parker, J., & Burkhard, B. (2016). *Irish wildlife manuals, national ecosystem and ecosystem service mapping pilot for a suite of prioritised services – appendices*. Ireland: National Parks and Wildlife Service, Department of Arts, Heritage, Regional, Rural and Gaeltacht Affairs.

Pavlík, M., Lukáčik, I., & Šulek, J. (2015). Rozklad odpadovej dendromasy drevokaznými hubami na modelových plochách v Arboréte Borová hora. In *Abstracts of Dendroflora of Central Europe – utilization of knowledge in research, education and practice*. Technical University in Zvolen, Zvolen, 10–11 June, 2015.

Pechoušková, V. (2006). *Modelování mělkých sesuvu a eroze v prostředí GIS GRASS*. Dissertation, Palacký University Olomouc.

Pérez-Soba, M., Harrison, P. A., Smith, A. C., Simpson, G., Uiterwijk, M., Miguel Ayala, L, Archaux, F., Erős, T., Fabrega Domenech, N., György, Á. I., Haines-Young, R., Li, S., Lommelen, E., Meiresonne, L., Mononen, L., Stange, E., Turkelboom, F., Veerkamp, C., & Wyllie De Echeverria, V. (2015). *Database and operational classification system of ecosystem service – natural capital relationships*. EU FP7 OpenNESS Project Deliverable 3.1. European Commission. http://www.openness-project.eu/sites/default/files/OpenNESS_D3.1_Final.pdf. Accessed 6 Apr 2019.

Perni, Á., Martínez-Paz, J., & Martínez-Carrasco, F. (2012). Social preferences and economic valuation for water quality and river restoration: The Segura River, Spain. *Water and Environment Journal, 26*(2). https://doi.org/10.1111/j.1747-6593.2011.00286.x.

Pielou, E. C. (1975). *Ecological diversity*. New York: Wiley.

Potts, S. G., Imperatriz-Fonseca, V., Ngo, H. T., Aizen, M. A., Biesmeijer, J. C., Breeze, T. D., Dicks, L. V., Garibaldi, L. A., Hill, R., Settele, J., & Vanbergen, A. J. (2016). Safeguarding pollinators and their values to human well-being. *Nature, 540*, 220–229. https://doi.org/10.1038/nature20588.

Považan, R., Getzner, M., & Kadlečík, J. (2014a). Hodnotenie ekosystémových služieb v chránených územiach Karpát so zameraním na Slovensko – Príručka pre rýchle hodnotenie. *Quaestiones Rerum Naturalium, 1*(2), 7–44.

Považan, R., Getzner, M., & Švajda, J. (2014b). Value of ecosystem services in mountain national park. Case study of Veľká Fatra National Park (Slovakia). *Polish Journal of Environmental Studies, 23*(5), 1699–1710.

Preston, S. M., & Raudsepp-Hearne, C. (2017). *Ecosystem service toolkit: Completing and using ecosystem service assessment for decision-making: An interdisciplinary toolkit for managers and analysts. Federal, Provincial, and Territorial Governments of Canada, Ottawa*. Ottawa: Environment and Climate Change Canada Enquiry Centre.

Rabe, S.-E., Koellner, T., Marzelli, S., Schumacher, P., & Gret-Regamey, A. (2016). National ecosystem services mapping at multiple scales – the German exemplar. *Ecological Indicators, 70*, 357–372. https://doi.org/10.1016/j.ecolind.2016.05.043.

Raudsepp-Hearne, C., Peterson, G. D., & Bennett, E. M. (2010). Ecosystem service bundles for analyzing tradeoffs in diverse landscapes. *Proceedings of the National Academy of Sciences, 107*(11), 5242–5247. https://doi.org/10.1073/pnas.0907284107.

Ricketts, T. H., Daily, G. C., Ehrlich, P. R., & Michener, C. D. (2004). Economic value of tropical forest to coffee production. *Proceedings of the National Academy of Sciences, 101*(34), 12579–12582. https://doi.org/10.1073/pnas.0405147101.

Ricketts, T. H., Regetz, J., Steffan-Dewenter, I., Cunningham, S. A., Kremen, C., Bogdanski, A., Gemmill-Herren, B., Greenleaf, S. S., Klein, A. M., Mayfield, M. M., Morandin, L. A., Ochieng, A., & Viana, B. F. (2008). Landscape effects on crop pollination services: Are there general patterns? *Ecology Letters, 57*(4), 499. https://doi.org/10.1111/j.1461-0248.2008.01157.x.

Rodríguez, J. P., Beard, T. D., Bennett, E. M., Cumming, G. S., Cork, S., Agard, J., Dobson, A. P., & Peterson, G. D. (2006). Trade-offs across space, time, and ecosystem services. *Ecology and Society, 11*(1), 28. http://www.ecologyandsociety.org/vol11/iss1/art28/.

ROTAP. (2012). *Review of transboundary air pollution: Acidification, eutrophication, ground level ozone and heavy metals in the UK*. http://www.rotap.ceh.ac.uk/files/RoTAP%20Summary%20report_0.pdf. Accessed 3 Apr 2019.

Royal Society. (2008). *Ground level ozone in the 21st century: Future trends, impacts and policy implications* (Science Policy report 15/18). London: The Royal Society. https://royalsociety.org/~/media/royal_society_content/policy/publications/2008/7925.pdf. Accessed 4 Feb 2019.

Rusch, A., Chaplin-Kramer, R., Gardiner, M. M., Hawro, V., Holland, J., Landis, D., Thies, C., Tscharntke, T., Weisser, W. W., Winqvist, C., Woltz, M., & Bommarco, R. (2016). Agricultural landscape simplification reduces natural pest control: A quantitative synthesis. *Agriculture, Ecosystems and Environment, 221*(1), 198–204. https://doi.org/10.1016/j.agee.2016.01.039.

Šabo, M., Minár, J., Holec, J., & Žiak, M. (2012). Syntetické hodnotenie vybraných prírodných hrozieb v oblasti Západných Karpát – prvé priblíženie. *Geografický časopis, 64*(4), 335–355.

Sandhu, H. S., Wratten, S. D., & Cullen, R. (2010). The role of supporting ecosystem services in conventional and organic arable farmland. *Ecological Complexity, 7*, 302–310. https://doi.org/10.1016/j.ecocom.2010.04.006.

Santos, T., Nogueira, M., & Vasco, A. (2016). Recreational activities in urban parks: Spatial interactions among users. *Journal of Outdoor Recreation and Tourism, 15*, 1–9. https://doi.org/10.1016/j.jort.2016.06.001.

Schröter, M., Bonn, A., Klotz, S., Seppelt, R., & Baessler, C. (2019). *Atlas of ecosystem services. Drivers, risks, and societal responses*. Cham: Springer.

Šefferová Stanová, V., & Galvánková, J. (Eds.). (2015). *Monitoring of plants and habitats of Community interest in the Slovak Republic. Results and assessment in the period of 2013–2015*. Banská Bystrica: State nature conservancy of the Slovak Republic.

Seják, J., & Dejmal, I. (2003). *Hodnocení a oceňování biotopů České republiky*. Praha: Český ekologický ústav.

Seják, J., Cudlín, P., Pokorný, J., Zapletal, M., Petříček, V., Guth, J., Chuman, T., Romportl, D., Skořepová, I., Vacek, V., Vyskot, I., Černý, K., Hesslerová, P., Burešová, R., Prokopová, M., Plch, R., Engstová, B., & Stará, L. (2010). *Hodnocení funkcí a služeb ekosystémů České republiky*. Ústí nad Labem: Jan Evangelista Purkyně University in Ústí nad Labem.

Shashua-Bar, L., Potcher, O., Bitan, A., Boltansky, D., & Yaakov, Y. (2010). Microclimate modelling of street tree species effects within the varied urban morphology in the Mediterranean city of Tel Aviv, Israel. *International Journal of Climatology, 30*(1), 44–57. https://doi.org/10.1002/joc.1869.

Šinka, K., Muchová, Z., & Konc, Ľ. (2013). *Aplikácie geografických informačných systémov v pozemkových úpravách*. Nitra: Slovak Academy of Sciences.

Skalský, R., Pavledna, P., Barančíková, G., Koco, Š., Barka, I., Tarasovičová, Z., & Makovníková, J. (2017). *Odhad zásob organického uhlíka v povrchovej vrstve pôd Slovenska*. Bratislava: Research Institute of Soil Science and Soil Protection.

Sláviková, D. (1987). *Ochrana rozptýlenej zelene v krajine. Metodicko-námetová príručka*. Bratislava: SZOPK.

Smelík, L. (2016). Analýza změn odtokových poměrů pro Českou republiku. *Vodohospodářské technicko-ekonomické informace, 58*(4), 7–12.

Smith, P., et al. (2011). Regulating services. In L. Davies et al. (Eds.), *UK National ecosystem assessment. Understanding nature's value to society* (Technical Report) (pp. 535–596). Cambridge: UNEP-WCMC.

Smith, P., Ashmore, M. R., Black, H. I., Burgess, P. J., Evans, C. D., Quine, T. A., Thomson, A. M., Hicks, K., & Harriet, G. (2012). The role of ecosystems and their management in regulating climate, and soil, water and air quality. *Journal of Applied Ecology, 50*(4), 812–829. https://doi.org/10.1111/1365-2664.12016.

Solín, Ľ. (2003). Koncepcia regionálnej hydrogeografie Slovenska. *Geografický časopis, 55*(2), 125–139.

Solín, Ľ. (2011). Regionálna variabilita povodňovej hrozby malých povodí na Slovensku. *Geografický časopis, 63*(1), 29–52.

Solín, Ľ. (2017). Flood Hazard in a mountainous region of Slovakia. In A. M. Negm & M. Zeleňáková (Eds.), *Water resources in Slovakia: Part II – Climate change, drought and floods* (Vol. 70, pp. 147–172). Cham: Springer. https://doi.org/10.1007/698_2017_172.

Solín, Ľ., Skubinčan, P., & Madajová, M. (2016). A preliminary flood-risk assessment of municipalities located in headwater basins of Slovakia based on an integrated approach. In C. A. Brebbia (Ed.), *WIT transactions on information and communication technologies* (Vol. 47, pp. 61–72). Southampton: WIT Press. https://doi.org/10.2495/RISK140061.

Špulerová, J. (2007). Nelesná vegetácia a jej hodnotenie pre potreby ochrany prírody. In *Abstracts of zo 4. Študentskej vedeckej konferencie: Ekológia a Environmentalistika*. Technical University in Zvolen.

Špulerová, J., Dobrovodská, M., Štefunková, D., Bača, A., & Lieskovský, J. (2014). *Biodiversity of traditional agricultural landscapes in Slovakia and their threats. Biocultural landscapes*. Dordrecht: Springer.

Špulerová, J., Štefunková, D., Dobrovodská, M., et al. (2017). *Historické štruktúry poľnohospodárskej krajiny Slovenska*. Bratislava: VEDA.

Špulerová, J., Petrovič, F., Mederly, P., Mojses, M., & Izakovičová, Z. (2018). Contribution of traditional farming to ecosystem services provision: Case studies from Slovakia. *Land, 7*(2), 74. https://doi.org/10.3390/land7020074.

Stanová, V., Valachovič, M. (2002). Katalóg Biotopov Slovenska. Bratislava: DAPHNE – Inštitút aplikovanej ekológie.

Stevens, M., Demolder, H., Jacobs, S., Michels, H., Schneiders, A., Simoens, I., Spanhove, T., Van Gossum, P., Van Reeth, W., & Peymen, J. (2015). *Flanders regional ecosystem assessment – State and trends synthesis report*. https://pureportal.inbo.be/portal/files/9004761/Stevens_etal_2015_FlandersRegionalEcosystemAssessment_State_Trends.pdf. Accessed 10 Apr 2019.

Streďanský, J., Dobák, D., Sollár, M., & Kliment, M. (2005). Súčasné spôsoby určovania intenzity veternej erózie v SR. In J. Rožnovský & T. Litschmann (Eds.), *Bioklimatologie současnosti a budoucnostipp*. Křtiny: Sborník abstraktů.

Supuka, J. (1998). Vegetačné formácie ako nástroj tvorby krajiny. *Životné prostredie, 32*(5), 229–232.

Sutton, M. A., Howard, C. M., Erisman, J. W., Billen, G., Bleeker, A., Grennfelt, P., & Grinsven, B. (2011). *The European nitrogen assessment: Sources, effects and policy perspectives*. http://assets.cambridge.org/97811070/06126/frontmatter/9781107006126_frontmatter.pdf. Accessed 10 Apr 2019.

Takács, Á., Kiss, M., & Gulyás, Á. (2014). Some aspects of indicator development for mapping microclimate regulation ecosystem service of urban tree stands. *Acta Climatologica et Chorologica, 47–48*, 99–108.

Tallis, H. T., Ricketts, T. H., Guerry, A. D., Nelson, E., Ennaanay, D., Wolny, S., Olwero, N., Vigerstol, K., Pennington, D., Mendoza, G., Aukema, J., Foster, J., Forrest, J., Cameron, D., Lonsdorf, E., Kennedy, C., Verutes, G., Kim, C., Guannel, G., Papenfus, M., Toft, J., Marsik, M., & Bernhardt, J. (2011). *InVEST 2.0 beta user's guide*. Stanford: The Natural Capital Project.

Termansen, M., Konrad, M., Levin, G., Hasler, B., Thorsen, B. J., Aslam, U., BOjesen, M., Hedemark Lundhede, T., Panduro, T. E., Estrup Andersen, H., & Strange, N. (2017). *Udvikling og afprøvning af metode til modellering af økosystemtjenester og biodiversitetsindikatorer –*

med henblik på kortlægning af synergier og konfl ikter ved arealtiltag. Aarhus Universitet, DCE – Nationalt Center for Miljø og Energi.

Tscharntke, T., Klein, A. M., Kruess, A., Steffan-Dewenter, I., & Thies, C. (2005). Landscape perspectives on agricultural intensification and biodiversity – Ecosystem service management. *Ecology Letters, 8*(8), 857. https://doi.org/10.1111/j.1461-0248.2005.00782.x.

UK NEA. (2011). *The UK national ecosystem assessment. Understanding nature's value to society* (Technical Report). Cambridge: UNEP-WCMC.

UNEP WCMC. (2011). *Approach for reporting on ecosystem services – UNEP-WCMC*. https://www.globalreporting.org/resourcelibrary/approach-for-reporting-on-ecosystem-services.pdf. Accessed 10 Apr 2019.

Vačkář, D., Frélichová, J., Lorencová, E., Pártl, A., Harmáčková, Z., & Loučková, B. (2014). *Metodologický rámec integrovaného hodnocení ekosystémových služeb v České republice*. http://www.ecosystemservices.cz/cs/metodologicky-ramec-integrovaneho-hodnoceni-ekosystemovych-sluzeb-v-ceske-republice. Accessed 13 Apr 2019.

Vanbergen, A. J., Baude, M., Biesmeijer, J. C., Britton, N. F., Brown, M. J. F., Brown, M., Bryden, J., Budge, G. E., Bull, J. C., Carvell, C., Challinor, A. J., Connolly, C. N., Evans, D. J., Feil, E. J., Garratt, M. P., Greco, M. K., Heard, M. S., Jansen, V. A. A., Keeling, M. J., Kunis, W. E., Marris, G. C., Memmott, J., Murray, J. T., Nicolson, S. W., Osborne, J. L., Paxton, R. J., Pirk, C. W. W., Polce, C., Potts, S. G., Priest, N. K., Raine, N. E., Roberts, S., Ryabov, E. V., Shafir, S., Shirley, M. D. F., Simpson, S. J., Stevenson, P. C., Stone, G. N., Termansen, M., & Wright, G. A. (2013). Threats to an ecosystem service: Pressures on pollinators. *Frontiers in Ecology and the Environment, 11*(5), 251. https://doi.org/10.1890/120126.

Vihervaara, P., Kumpula, T., Tanskanen, A., & Burkhard, B. (2010). Ecosystem services – A tool for sustainable management of human–environment systems. Case study Finnish Forest Lapland. *Ecological Complexity, 7*(3), 410–420. https://doi.org/10.1016/j.ecocom.2009.12.002.

Vihervaara, P., Kumpula, T., Ruokolainen, A., Tanskanen, A., & Burkhard, B. (2012). The use of detailed biotope data for linking biodiversity with ecosystem services in Finland. *International Journal of Biodiversity Science, Ecosystem Services and Management, 8*(1–2), 169–185. https://doi.org/10.1080/21513732.2012.686120.

Vilček, J. (2014). Mapovanie a hodnotenie environmentálnych funkcií poľnohospodárskych pôd Slovenska. *Geografický časopis, 66*(3), 287–304.

Vranic, P., Zhiyanski, M., & Milutinovic, S. (2016). A conceptual framework for linking urban green lands ecosystem services with planning and design tools for amelioration of microclimate. *Journal of Integrative Environmental Science, 13*(2–4), 129–143. https://doi.org/10.1080/1943815X.2016.1201516.

Žiak, M. (2012). *Lavínová hrozba, bilancia energie a hmoty vo vysokohorskom prostredí*. Dissertation, Comenius University in Bratislava.

Zulian, G., et al. (2013). *ESTIMAP: Ecosystem services mapping at European scale*. Brussel: Institute for Environment and Sustainability, Joint Research Centre, European Commission.

Chapter 5
Cultural Ecosystem Services

Zita Izakovičová, Dagmar Štefunková, Jana Špulerová, Dominika Kaisová, Gréta Vrbičanová, Peter Mederly, František Petrovič, Matej Močko, Martina Turanovičová, Barbora Šatalová, Simona Gusejnov, Tomáš Kováč, Ján Černecký, and Viktória Ďuricová

Abstract The chapter provides the analysis and assessment of **three cultural ES**: *C1, Recreation and Tourism: Physical Use of Nature and Landscape; C2, Landscape Aesthetics: Aesthetic Values;* and *C3, Natural and Cultural Heritage: Intellectual and Scientific Values.* All ES are described in the unified structure: definition and brief characteristics, methods used for identification and assessment, main types of landscape and ecosystems providing given ES, the importance of ES in terms of nature and landscape protection and ES assessment for the territory of Slovakia. Spatial assessment is provided as a map of the landscape capacity for a given ES provision. For all ES, short conclusions and overview of input data for further assessment of the ES capacity, demand and flow are also given.

Z. Izakovičová (✉) · D. Štefunková · J. Špulerová · B. Šatalová
Institute of Landscape Ecology of the Slovak Academy of Sciences, Bratislava, Slovakia
e-mail: zita.izakovicova@savba.sk; dagmar.stefunkova@savba.sk; jana.spulerova@savba.sk; barbora.satalova@savba.sk

D. Kaisová · G. Vrbičanová · P. Mederly · F. Petrovič · M. Močko · M. Turanovičová
S. Gusejnov · T. Kováč
Faculty of Natural Sciences, Constantine the Philosopher University in Nitra, Nitra, Slovakia
e-mail: dominika.kaisova@ukf.sk; greta.vrbicanova@ukf.sk; pmederly@ukf.sk; fpetrovic@ukf.sk; matej.mocko@ukf.sk; martina.turanovicova@ukf.sk; simona.stasova@ukf.sk; tomas.kovac@ukf.sk

J. Černecký
Faculty of Natural Sciences, Constantine the Philosopher University in Nitra, Nitra, Slovakia

State Nature Conservancy of the Slovak Republic, Banská Bystrica, Slovakia

Institute of Landscape Ecology of the Slovak Academy of Sciences, Bratislava, Slovakia
e-mail: jan.cernecky@sopsr.sk

V. Ďuricová
State Nature Conservancy of the Slovak Republic, Banská Bystrica, Slovakia

Faculty of Natural Sciences, Matej Bel University, Banská Bystrica, Slovakia
e-mail: viktoria.duricova@sopsr.sk

P. Mederly, J. Černecký (eds.), *A Catalogue of Ecosystem Services in Slovakia*,
https://doi.org/10.1007/978-3-030-46508-7_5

5.1 Recreation and Tourism: Physical Use of Nature and Landscape (C1)

5.1.1 Definition and Brief Characteristics of ES

Natural ecosystems provide us with almost endless opportunities for spiritual enrichment, mental development and leisure. Nature is an important source of inspiration for science, culture and art and provides many opportunities for education and research (Gallagher 1993).

Two views are emerging when defining recreation in relation to the ES, in particular to cultural ES. The first considers recreation as an ecosystem service, while the second refers to recreation as a benefit. This is based on an inconsistent understanding of service and benefit terms in the ES context. Boyd and Banzhaf (2007) and Fisher et al. (2009) are in favour of claiming that recreation is only a benefit consisting of multiple inputs (human, social and economic capital) and the ES, which can contribute to the production of recreational benefits and can be present in a number of ecological elements such as forests and meadows.

On the other hand, there are scientists who see recreation as an ES and define it as *the restoration and stimulation of the human body and soul through exploration and interaction with living organisms in their natural environment* (Beaumont et al. (2007)) or *the pleasure which people obtain from natural and cultural ecosystems* (Nahuelhual et al. 2013).

Recreation seen as an ES is one of the cultural ES where ecosystems provide non-material benefits to people (Lankia et al. 2015), and we understand it in terms of user movement, as the service demand is closely related to the presence of people in ecosystems (Nahuelhual et al. 2013). People choose their place of leisure based on the state of the natural and cultivated landscape in the given area (MEA 2005). The role which the ecosystems have in maintaining mental or physical health is highly recognized, despite the difficulties in its measurement (TEEB 2013).

5.1.2 Methods Used to Assess and Identify ES

The value of cultural ES varies from one individual to another. Sociocultural assessment is used most often for the assessment of recreation and tourism. This includes qualitative methods, for example documents/photos analysis from Panoramio that was used in Israel's national study (Lotan et al. 2018) and from Flickr social network in Luxembourg (Becerra-Jurado et al. 2016), or quantitative methods (e.g.

questionnaires) used in a case study in Slovakia (Považan et al. 2015) or Great Britain (Schmidt et al. 2016).

As part of the economic assessment, the *contingent valuation* is preferred: willingness to pay/willingness to accept (WTP/WTA) is understood as the hypothetical cost of entering protected areas and costs of accommodation, meals, fuel or tickets in public transport. This method was applied in case studies in Slovakia (OZ Pronatur 2014; Považan 2013; Považan et al. 2014a; Getzner 2009; Fűzyová et al. 2009). Travel cost method (real consumer costs associated with accommodation, food, transport and entrance fees) was used in Italian protected areas (Schirpke et al. 2018). The economic assessment of recreation by means of value transfer has been applied by Frélichová et al. (2014) in the CR. This method is cost-effective but is susceptible to generalization errors for a number of input variables.

In some cases, ecological (biophysical) methods, such as *biodiversity indicator development*, are also used to assess recreation and tourism. These methods analyze environmental variables to indicate biodiversity status and changes. According to SEA, the key indicators for tourism in Slovakia include, for example, tourism destinations, erosion caused by tourism or a number of protected areas threatened by tourism (available online: www.enviroportal.sk/indikatory). The use of models can significantly contribute to spatial data representation; such an approach was proposed in the CR (Vačkář et al. 2014).

Other methods are also used in various studies, which can generally be used to assess the cultural ES. Many of them are described in the assessment of the next ES C2.

5.1.3 The Main Types of Landscape and Ecosystems Which Provide ES

Every natural or seminatural ecosystem (landscape unit) can provide several cultural ES. From a societal and cultural point of view, grasslands (meadows and pastures) help maintain the viability of rural communities as an important source of employment, improving rural tourism and recreation (Kemp and Michalk 2007). They offer suitable conditions for ecotourism and education (nature trails, hiking with expert guide).

In a natural as well as human-modified landscape, the rivers, water bodies and other water elements (e.g. fountains or small ponds in recreational areas of cities or in private gardens) play an important aesthetic value. They are also intensively used for recreational purposes (bathing, boating, rafting, canoeing, sport fishing, photography or ecotourism).

Mountain ecosystems provide countless cultural ES. Along with the rich dissection of the relief, they have year-round importance for the development of recreation and tourism. They are particularly important in terms of winter sports development; in summer they are used mainly for hiking, forest fruit harvesting and so on

Fig. 5.1 The recreational use of landscape is one of the most important functions of mountain ecosystems (mountain hut – Chata pod Borišovom, Veľká Fatra), author: D. Kaisová

(Fig. 5.1). The attractiveness of these sites is increasing with the development of ecotourism, which aims at education and nature protection (Vandewalle et al. 2009).

In addition to its primary production function, the agricultural landscape has been developing as a tourist destination in recent years. It offers opportunities to learn about the historical, cultural and natural potential of the landscape. An example of the intersection of agriculture and tourism is agro-tourism (Palkechová and Kozáková 2015).

Geological formations and geomorphological phenomena represent a unique aesthetic element of the landscape in terms of cultural ES, which form the main attraction for establishing educational trails or for active leisure – climbing or speleotourism (Hanley et al. 2002).

The phenomenon of a healthy lifestyle also brings forward the active use of spaces of urban parks and other urban vegetation, especially for physical activities such as running, walking, skating or cycling. Maintained greenery, water areas, playgrounds and other green space equipment create suitable conditions for recreation, which, thanks to the relatively simple accessibility (in the city), is widely used for spending leisure time (Santos et al. 2016).

5.1.4 Importance of ES in Terms of Nature and Landscape Protection in Slovakia

Recreation and tourism are associated with different types of landscapes. However, it is important for all types of landscape to have something to offer to visitors, whether it be natural wealth or cultural-historical sites. That is why many

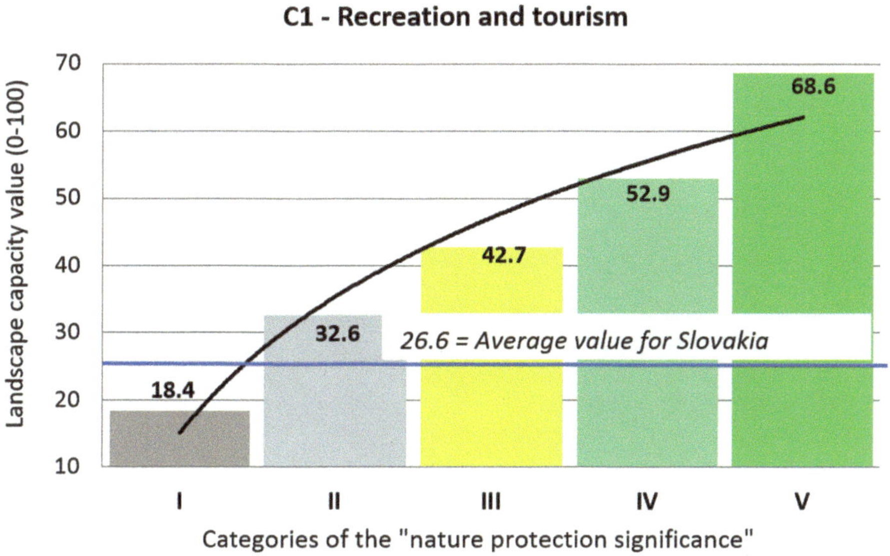

Fig. 5.2 The relationship between ecosystem service C1 and the significance of the territory of Slovakia in terms of nature and landscape protection

holidaymakers seek recreation and tourism in areas which are unique with their species richness or landscape structure. These areas are often subject to some level of territorial protection. Figure 5.2 shows a direct correlation between the significance of the territory of Slovakia in terms of nature and landscape protection and the landscape's capacity for recreational ES.

Visitor's activities in protected areas are regulated by Act No. 543/2002 on Nature and Landscape Protection, which sets limits and restrictions on the movement of visitors from the second degree of protection. Increasing the degree of protection also increases the degree of limitations for visitors. These limitations aim to provide sufficient species and territorial protection for the most valuable parts of the landscape. Unregulated movement of visitors could result in irreversible disturbances of ecological balance in protected areas, for example, disturbing game at the time of breeding and bringing their young out or by trampling on the habitats of protected plant species. The increase of visitors can lead to loss of soil on the hiking trails, the widening of trails and secondary trails and changes in the species composition of vegetation. In addition to maintenance, the solution could be in the form of regulation of number of visitors. However, the situation is complicated by complex legislation, property ownership relationships, demanding implementation and non-understanding from the general public (Vološčuk et al. 2016; Piscová et al. 2018).

On the other hand, recreation and tourism can help preserve valuable areas, for example, with entry fees, from which various educational and training events can be financed within protected areas or to cover the costs of protected area management. By visiting such areas, visitors can learn about fauna, flora, important geological features and cultural-historical objects through educational trails, which contributes to public education about nature and landscape protection (Fig. 5.3).

Fig. 5.3 Flagship species (interesting for ordinary people who know them) are a form of enrichment of tourism and nature observation but also an essential component of ecosystem functioning, author: J. Černecký

The integration of soft forms of tourism, which include ecotourism, agro-tourism and rural tourism (Pásková 2008, Pourová 2000), into the decision-making process of municipalities and nature and landscape protection administration bodies plays a particularly important role because some of the cultural ES is very difficult, even impossible, to replace. These include services directly affected by intensive, mass tourism (e.g. aesthetics, natural and cultural heritage).

5.1.5 ES Assessment for the Territory of Slovakia

To calculate the value of recreation and tourism in a specific territory (e.g. protected areas), it is possible to apply procedures according to the manual (Považan et al. 2014b); however, this approach is currently not applicable at a national level due to the absence of the necessary data.

The travel cost method can be used for economic evaluation. The V_{RT} value of recreation/tourism is based on contingent assessment and visitor statistics. The average cost per person per day and the average length of a visitor's stay are usually assessed in the form of a visit survey. *Total travel costs* can be calculated based on the following formula (Považan et al. 2014b): $TC_a = N_a * D_i (TC_{i,1} + TC_{i,2}) * M_a$, where TC_a means the total travel costs of visitors per year a, N_a is the number of visitors per year a, D_i is the average length of stay of visitors i, $TC_{i,1}$ are travel costs in terms of purely travel costs (transport), $TC_{i,2}$ means other costs associated with the visit (e.g. expenses for accommodation or souvenirs) and M_a is the average number of visitors who only come to see the territory.

For a clear assessment of recreation and tourism as an ES at national level, it is important to allocate areas which provide recreational opportunities while taking into account territorial and species protection, which partially represents limits. In this step, it is necessary to use all suitable spatial information (e.g. maps of landscape structure and use, important natural and cultural-historical sites), so that we can allocate suitable and attractive places for recreation. This data may be supplemented by statistical data showing the number of accommodation and restaurant facilities and the number of beds and parking spaces (at municipal or district level).

It is also important to include in the assessment the environmental limits, such as environmental quality assessment and assessment of selected negative factors, for example mining areas, industrial sites, damaged areas, polluted air, noise and contaminated watercourses. Climate data (number of sunny days per year, number of days with snow cover) and hydrological data may be also included in the assessment. Input data are shown in Table 5.1.

Table 5.1 Input data for assessment of ES recreation and tourism – physical use of nature and landscape

Input data/ES	C1 – Recreation and tourism
Capacity	Map of locations of European network NATURA 2000 (SAC and SPA)
	Map of protected areas (small scale, large area)
	Map of important habitats and habitats of protected plant and animal species
	Map of cultural and historical monuments and reservations
	Map of hiking trails, bike paths, ski resorts, cross-country skiing and running trails
	Statistical data on the number of accommodation facilities per unit area
	Statistical data on the number of restaurant facilities per unit area
	Number of beds (for leisure) per unit area
	Map of hunting and fishing grounds
	Assessment matrix – the relationship between CLS units and recreational ES
	Accessibility of territories from cities (e.g. from regional centres)
	Data about environmental limits (e.g. environmental quality)
	Climatological and hydrological data – temperatures, solar radiation, cloud cover, snow cover
Demand	Map of the current landscape structure – significant and attractive elements of CLS
	Substrate and relief map – attractive shapes and forms of relief and geological environment
	Data on population distribution – urban settlements and densely populated areas
	Map of tourism-related infrastructure (location of hotels, restaurants, car parks, recreational areas)
Flow	Photos on social networks – number of visits and attractiveness of selected areas
	Visitor's statistics for selected territories (e.g. number of overnight stays per unit area per time; number of cave tickets sold; tickets for cultural-historical monuments per unit of time; number of fishing permits per unit area per time; amount of sport equipment rented – bicycles, skis, scooters – per unit of area per time; number of ski passes sold and lift tickets per unit of time)

From the point of view of the spatial expression of the assessed landscape capacity for provision of this ES, it is clear that the most favourable values are typical for the alpine and mountain areas of Slovakia with significant natural conservation values (Fig. 5.4) – especially the areas Vysoké Tatry, Nízke Tatry, Malá Fatra, Poľana and Malé Karpaty. Other important areas include larger areas of hydric ecosystems – especially the area of Podunajsko, the Latorica area and large water reservoirs. On the contrary, the lowest values of recreational potential are typical for large-scale lowlands and basins – especially the Východoslovenská nížina lowland, parts of the Podunajská nížina lowland and Juhoslovenská kotlina basin (Fig. 5.5).

Slovakia has a rich potential for the provision of recreational services and tourism. These services are used by domestic visitors, but foreign tourists are also significantly involved in using this potential. The demand for recreation and tourism is strongly dependent on subjective factors – the various interests of visitors. Demand for recreation and tourism is also influenced by many factors characterizing human potential – social status, education, place of residence, income and so on. Low-income groups do not have the opportunity to develop recreational activities and tourism, as these activities are often associated with certain financial expenses. A similarly important role in the demand for this ES is also played by the place of

Fig. 5.4 Turňa Castle Hill is part of the Slovenský kras National Park, which attracts tourists every year, author: J. Černecký

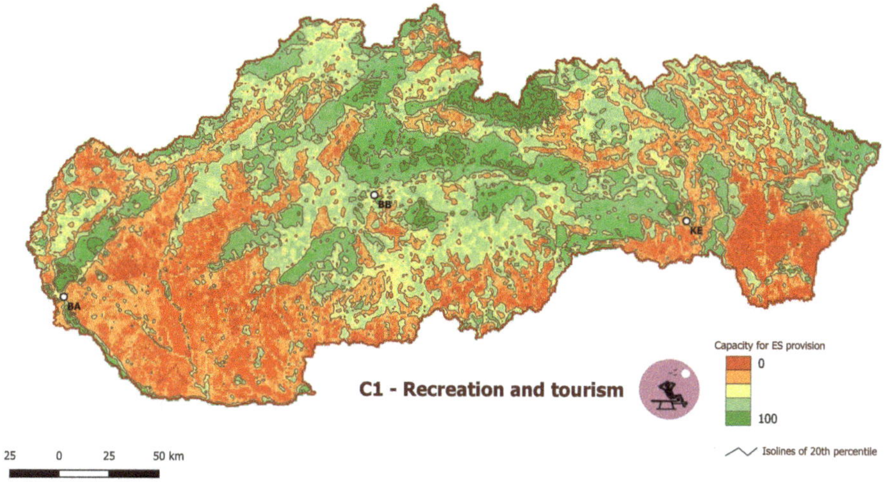

Fig. 5.5 Capacity of the landscape to provide ES recreation and tourism – physical use of nature and landscape

residence, with the urban population having more interest in tourist activities and recreation. The rural population shows less interest in tourism and recreation, as its life is more connected with nature in the form of relaxation and work in the backyard gardens, the performance of self-supply food production and the like. Backyard gardens, orchards and vineyards are also often part of rural homes, which in many cases replace nature.

The real use of ES recreation and tourism – physical use of nature and landscape – is quite difficult to assess objectively, as number of visitors is not monitored in all locations. Suitable indicators for monitoring the use of this ES include statistical data on number of visits to selected areas (e.g. number of overnight stays per unit area per time; number of cave tickets sold; tickets for cultural-historical monuments; number of fishing permits; amount of sports equipment rented – bicycles, skis, scooters; number of ski passes sold and lift tickets; the number of tickets for swimming pools and aquaparks). Demand for free tourism, forest fruit harvesting, walks, etc. is harder to assess in cases where a number of visits is not recorded. Often, these activities are also linked to climatic conditions. Interest in tourism and water sports is also increasing with increasing temperatures and changing climate. Similarly, winter sports are limited by the duration and quality of snow cover. Therefore, the most suitable method to determine the demand and real use of this ES includes the questionnaire methods and various other ways to determine the preferences of residents and visitors.

5.2 Landscape Aesthetics: Aesthetic Values (C2)

5.2.1 Definition and Brief Characteristics of ES

The assessment of the aesthetic values of the landscape and the perception of its beauty is based on an interdisciplinary approach to landscape research, which understands the landscape as a material system with its measurable objects and processes (ecosystem approach), but also evaluates the intangible dimensions of the landscape characteristic of a holistic understanding of the landscape. Landscape perception has a physiological and psychological aspect. The landscape is perceived on the human scale. It provides a view from horizontal observation points (Oťaheľ 1999, 2003) determined by environmental properties and the observer.

Depending on the field of science, approach and method, the current terminology base in Slovakia includes most frequently the following terms – *aesthetic quality of the landscape, landscape image, landscape scenery, visual quality of the landscape, landscape character or characteristic landscape appearance* (Mišíková 2002; Wöbse 1991; Drdoš 1995; Oťaheľ 2003; Štefunková 2004). In Slovakia, the preferred term is the *characteristic landscape appearance* (CLA). The CLA identification and assessment methodology published (Jančura et al. 2010) defines CLA as a set of characteristics which distinguish one landscape from any other.

The visual attractiveness, landscape image and landscape scenery, as well as other cultural ES, are among the ES with the most prominent influence of subjectivity in the assessment process. On the other hand, these ES have their value and play an important role in encouraging public support for ecosystem protection (Daniel et al. 2012). The visual quality of landscape and ecosystems affects the quality of life or the aesthetic enjoyment of people from ecosystem and landscape observation (Burkhard et al. 2014). It also has a special relationship to other cultural ecosystem services, as it greatly supports their value – strong aesthetic effect of ecosystems determines the realization of recreational activities and tourism; strong sensual (aesthetic) experience positively affects the assignment of moral and spiritual values to the perceived place by residents or visitors.

Service beneficiaries – residents, visitors and stakeholders – have different individual and collective value profiles and requirements and the nature of relationships to the perceived image and scenery (Zube et al. 1982). The properties of ecosystems which are measurable and suitable for assessing their attractiveness are filtered through observer personality, psychological and physiological state and cognitive schemes in the process of landscape perception. An important factor in the perception of the landscape also includes the properties of the environment, such as the current and seasonal climatic conditions and the position and distance from the observed scenery (Štefunková 2004; Nohl 1991; Krause 1991). The psychophysical paradigm in the research of the perception of the landscape is based on the link between the material/spatial arrangement of the landscape (the elements comprising it) and the observer's assessment (Zube et al. 1982).

5.2.2 Methods Used to Assess and Identify ES

There are several concepts in the current assessment of this type of ES. An *expert-based approach* (Zube et al. 1982) is an assessment of the visual characteristics of selected types of landscape or its parts by trained professionals. The biophysical properties of the landscape are transformed into formal features (e.g. lines, textures and colours) and the relationships between them (e.g. diversity, unity, harmony), by applying professional methods and procedures to evaluate the landscape's own aesthetic effect.

The essence of methods based on measuring the perception of the landscape (Daniel 2001) or subjective methods (Barčáková 2001) is that they focus on the assessment of the quality of the landscape by an observer – whether it is a tourist or a resident who permanently lives in the landscape or uses it in different ways. These behavioural approaches include, for example, structured and unstructured questionnaires, semantic differential and mental maps.

Combined approaches have been used after lower reliability has been confirmed in using one of the previous approaches. The approach of inventory and scenario assessment and visual landscape quality is based on parallel expert and behavioural assessment, which is subsequently compared and assessed for validity (Daniel 2001).

The aesthetics and beauty of ecosystems and landscapes are most often assessed by methods of indirect monetary evaluation (e.g. willingness to pay) or by nonmonetary quantitative assessments such as preferential assessment, number of visitors and psychometric scales (Daniel et al. 2012). Economic indicators such as property prices can be also used for the assessment (Milcu et al. 2013).

An example of the use of nonmonetary quantitative methods is the research of the visual quality of a part of the Tuscan landscape (Sottini et al. 2018), implemented in three steps: landscape classification into landscape mosaic types (from mostly urbanized mosaics to predominantly natural mosaics), research of respondent's perception (public) on the basis of photographs of these types of land sites and statistical analysis and assessment of sociological research results.

The capacity of the territory (ecosystems) to provide an aesthetic experience is often assessed through the physiognomic structure of the landscape, based on criteria directly derived from the structural and physiognomic characteristics of the landscape (e.g. vegetation cover, length of the borders) or indirectly derived criteria including an aspect of the expected impact on the personality of the perceiver (e.g. harmony, attractiveness, uniqueness). The US Department of the Interior Bureau of Land Management developed a guide for landscape visual quality assessment which uses seven indicators – relief, vegetation cover, colour, water features, adjacent scenery, scarcity and cultural transformation character (Brown 1994).

Research of viewshed analysis and the identification of scenic viewpoints and sceneries in the landscape are an important part of either normative or combined research. For example, Oťaheľ (1999) assessed the location of recreational facilities based on the assessment of the optimal view of the High Tatras. Štefunková and Cebecauer (2006) assessed the visual quality of selected areas by modelling the potential viewshed and visual dominance of the landscape in the GIS environment in combination with the assessment of the aesthetic quality of landscape features.

Analysis of photographs – landscape, its individual elements, structures, scenery through respondents (visitors or residents of the studied territory), or a specific view on which they are located – is also a frequently used method. If the photo is placed directly at the place where it was taken and the place is publicly accessible, the respondent directly expresses the demand for or benefit from the use of ES with its assessment.

Analysis of the so-called big data such as Panoramio photos (Giglio et al. 2019; Lieskovský et al. 2017) can involve a much larger number of respondents and thus carry out assessments at the national or continental level. It directly expresses the benefits of the visual perception of the given place or landscape. The results are strongly influenced by accessibility to scenic sites. Most photos are taken at a place where people live or go on vacation, but as the authors of the study comment, the demonstration of the use of and benefit from the ES is relevant.

In order to eliminate the level of subjectivity, the assessment of the visual (aesthetic) quality of the landscape performed by experts uses multicriterial decision analysis (MDCA) procedures to assess conflicting ES.

5.2.3 The Main Types of Landscape and Ecosystems Which Provide ES

There are a number of studies conducted with an assessment of the aesthetic/visual/scenic quality of the landscape in different scales and at different levels. Most of the time authors choose an area which is attractive to residents and visitors, whether it is a landscape heritage and traditional agrarian landscape, protected natural area, river floodplains and coastal landscapes or distinct geomorphological unit. On a local scale, selected landscape sceneries which represent a group of ecosystems in a specific composition and combination are most frequently assessed. Combinations of natural dominants with barrier-free foregrounds are attractive (such as a view of Vysoké Tatry from the grasslands of the Liptovská kotlina basin or the scenery of forests with a water surface or meadow in the foreground). The prerequisite for the real use of this ES is its valuation in the form of, for example, the high number of visitors in the city park, the river promenade, the location of the lookout towers, new residential areas, recreational areas and hiking trails. One can see a preference of certain types of landscape, which are most preferred by the general criteria and for which high visual-aesthetic quality (Fig. 5.6) has been proven by previous research. These types of landscapes are also highly valued by experts in the methods and classifications they design.

Fig. 5.6 The aesthetic landscape character of rural settlements in remote regions is becoming increasingly attractive for domestic and foreign visitors (Detvianske lazy), author: J. Černecký

The result of measurement of Corine Land Cover classes contribution to aesthetic experience in a selected region of Germany (Koschke et al. 2012) has shown that the maximum contribution – the highest potential – comes from mixed forests, transitional woodland/shrub, watercourses and water bodies, wetlands and deciduous forests. Significant values have also been achieved by the following classes: a mosaic of fields, meadows and permanent cultures, coniferous forests, pastures and agricultural areas with a significant proportion of natural vegetation.

In behavioural research by Lieskovský et al. (2017), who assessed the attractiveness of landscape types by the Panoramio photos, the most attractive landscape types were sub-mountain and mountain meadows and urbanized landscapes in the river basin and highland relief. These mainly included the highest-elevation tourist locations with panoramic views in the Vysoké Tatry and cultural heritage sites such as Spiš Castle, Oravský Podzámok and Červený Kláštor. Similar visitor preferences were shown by another study (Othman 2015), where respondents prefer the forest landscape, hilly terrain, and architectural heritage objects the most. The water bodies and waterfalls were not the most preferred here, which could have been caused by environmental pollution. In light of the above, the authors said that scenic beauty can be an indicator of the good environmental conditions of the considered landscape. In the assessment of forests, there are known efforts to assess the non-production functions of the forests, where, for example, forest parks, therapeutic (spa) forests and recreational forests with a specific structure, composition and representation of visually attractive tree species meet the highest criteria of visual attractiveness (Supuka and Vreštiak 1984). Locations significant in terms of ES provision include urban and suburban forests and parks and riversides with natural sceneries due to high number of visitors, which expresses the demand and benefit of the visual aesthetic effect of natural and seminatural ecosystems on humans.

5.2.4 *Importance of ES in Terms of Nature and Landscape Protection in Slovakia*

Proper setting and planning of a nature protection strategy in important habitats and visually highly attractive sites are necessary as the limits for the number of visitors of tourist trails in these territories are highly exceeded (Švajda 2009) and this damages significant habitats (Hrnčiarová et al. 2018). To wisely plan the number of visitors, it is necessary to redirect visitors to less vulnerable ecosystems or to provide them with a scenically attractive experience from viewing points located outside the threatened habitats. Above all, it is important to know the motivation of visitors to these sites (Švajda et al. 2018; Považan et al. 2015), as well as to know all the risks endangering the habitat status near to highly used tourist routes (Špulerová et al. 2014). Forests of protected areas do not differ from commercial forests if they do not have any sustainable management (Považan et al. 2014a). Although most forests in protected areas are not state property, protected areas administration organizations may provide comments on the forest management plans. However, as Švajda and Fenichel (2011) showed, effective ecological management of protected areas in Slovakia is hampered by the lack of authority of these administrations.

While a number of European countries have committed themselves to providing adequate funding to protected areas, Slovakia is still lagging behind, and ES benefits are being reduced compared to the potential economic benefits of intensive tourism development in protected areas, especially in national parks. Research in the three national parks of Slovakia has shown that they are all highly sought after and visited for recreation and tourism purposes. The aesthetic (visual) attractiveness of the landscape is an important condition for the motivation of tourists to visit such areas. Therefore, it is necessary to deal with the assessment of this ES within protected areas, to set the price of the service and subsequently achieve an effective consensus of service management with the owners, stakeholders and the public. A clear positive correlation between the significance of the territory in terms of nature and landscape protection and the capacity of the landscape to provide ES landscape aesthetics – aesthetic values – is documented in Fig. 5.7.

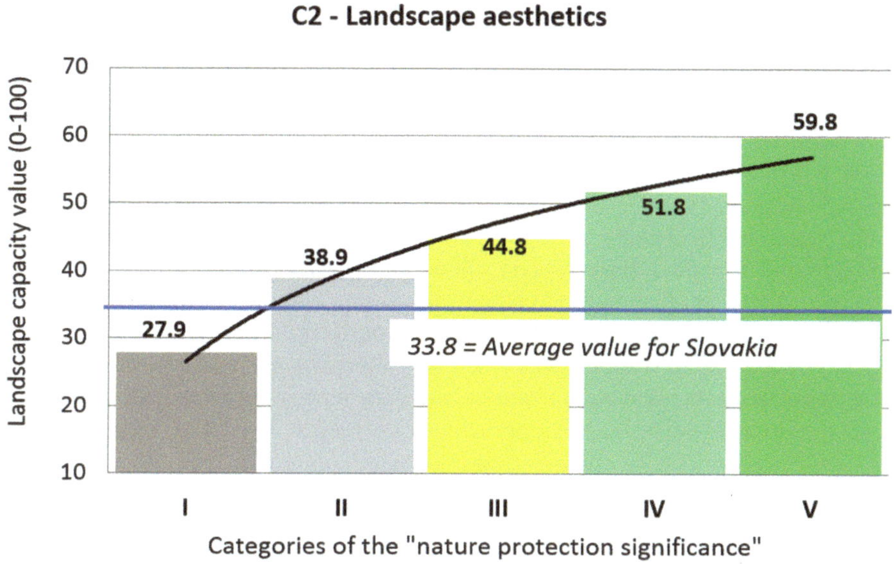

Fig. 5.7 The relationship between ecosystem service C2 and the significance of the territory of Slovakia in terms of nature and landscape protection

5.2.5 *ES Assessment for the Territory of Slovakia*

In the field of scientific assessment of the visual (aesthetic) quality of the landscape at the national level, we know only of a few examples. An older study on the socio-economic assessment of vegetation was published by Jurko (1990) – the assessment topic was the aesthetic significance of vegetation formations. Selected criteria included increasing the quality of the landscape by spatially dividing the areas, facilitating orientation and aesthetic experiences (colouring of flowering, foliage and fruit in autumn), shading or masking of negatively acting objects and the attractiveness of nature observation in terms of its diversity.

In more recent national assessments, the authors focused on viewshed analysis or the so-called big data analysis. Jakab and Petluš (2013) created a map of the potential visual exposure of the landscape relief of Slovakia in the GIS environment. The best category includes, for example, the important tourist points – Záruby on the main ridge of the Malé Karpaty, Zobor in the Tribeč mountains, Babia Hora in the Oravské Beskydy, Gerlachovský štít in Vysoké Tatry.

Regional and national surveys, in contrast to local studies, are based more on the assessments performed by experts, as it is difficult to implement such extensive research through respondents. The only behavioural-oriented national research based on Panoramio photographs was performed by Lieskovský et al. (2017).

In the research carried out by experts focusing not only on the potential but also on the demand and benefit of the perception of ecosystem beauty, it is recommended, in addition to identifying the visual (aesthetic) value of ecosystems, to analyze whether and to what extent these ecosystems can be perceived from sites with high number of tourists or residents. Such an assessment is already conceptually linked not only to the identification of tourism hotspots and their availability but also to the identification of all other transport nodes with a high concentration of movement or stay of people, such as highways and 1st-class roads in elevated positions, with a view on visually landscape elements. Also, the locations of spas, recreational houses and cottage areas are an indirect expression of the demand and benefit from perceiving the beauty of ecosystems in their surroundings. However, in such an assessment method, the aesthetic value may overlap with the value of tourism and recreation – another cultural ES.

Based on the available underlying databases, the assessment of the aesthetic (visual) quality of ecosystems was based on the importance of CLS (classification in terms of aesthetics and attractiveness), the occurrence of special landscape structures (traditional land use) and the attractiveness of relief. The following was chosen as supplementary criteria: quality of forest ecosystems (forest type and age), visual diversity of the landscape (dissection and slope inclination of the microbasin), significant aesthetic elements of the landscape, historical and cultural monuments, historical vegetation and other natural attractions (Fig. 5.8). The list of data used for the assessment of ES landscape aesthetics – aesthetic values – is shown in Table 5.2.

The spatial expression of the landscape's capacity to provide this ES (Fig. 5.9) highlights the importance of the diversity of the landscape structure and the diversity of natural conditions – the highest values are achieved in mountain areas (especially in regions of Vysoké Tatry, Nízke Tatry, Veľká Fatra, Malá Fatra, Slovenský raj, Slovenské Rudohorie, Strážovské vrchy, Javorníky, Štiavnické vrchy, Malé Karpaty). Lower mountain, submontane and basin areas mostly reach a moderate capacity. Finally, the lowest capacity values are typical for larger areas of lowlands and river basins.

Demand for ES landscape aesthetics – aesthetic values – depends on the interest of the population in beauty, aesthetics, visual quality of the landscape and so on. The perception of the visual quality of the landscape is influenced by a variety of factors, such as composition, structure, attractiveness, uniqueness and land management. An important role is played by the subjective factor – individual experience.

Fig. 5.8 An attractive view of the natural panorama of the Cigánka nature reserve in the Muránska Planina National Park, author: J. Černecký

Table 5.2 Input data for assessment of ES landscape aesthetics – aesthetic values

Input data/ ES	C2 – Landscape aesthetics – aesthetic values
Capacity	Map of current landscape structure – significant and attractive elements of CLS, reclassification according to aesthetic effect
	Land use – occurrence of specific cultural and historical landscapes (traditional method of use, dispersed settlement areas, wine-growing areas, mining landscape)
	Specific features of the landscape – historical parks, spas
	Forest ecosystems – reclassification by structure and age, type of vegetation
	Relief – attractive shapes and forms of relief
	Relief – dissection and slope of microbasins (visual diversity of the landscape)
Demand	Data on population distribution – urban settlements and densely populated areas
	Data on the number of visitors – municipalities, regions
	Infrastructure map (location of hotels, restaurants, car parks, recreational areas and so on)
Flow	Photos on social networks – number of visits and attractiveness of selected areas
	Statistical data on the number of visits to selected areas – as in the case of C1

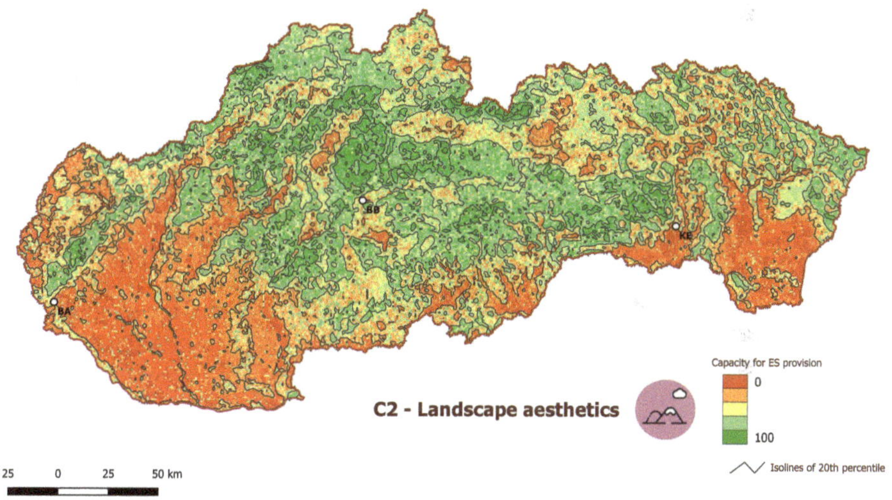

Fig. 5.9 Capacity of the landscape to provide ES landscape aesthetics – aesthetic values

Beauty, aesthetics and visual quality are perceived differently by individuals. One may find something very interesting, with someone else perceiving it as very common. Each of us prefers a different type of landscape and landscape character. Demand for this ES can therefore also be assessed from a number of perspectives – from the perspective of visitors (number of visitors), from the perspective of residents (preference of landscape types for housing), from the perspective of researchers (number of realized assessment studies) and from the perspective of artists (the number of works of art produced). Demand for this ES can also be linked to the demand for another cultural ES and cannot be clearly separated from them.

Assessment of the real use of this ES is very difficult, as subjectivity plays an important role here. In general, landscape types with high visual quality are freely available, in most cases without a fee, and so an objective tracking of number of visitors and their interest is not possible. For the assessment of real use, indirect methods may be used in particular, such as landscape scenarios based on photographs and various forms of sociological surveys – structured and unstructured questionnaires and personnel interviews. It is also possible to carry out a targeted counting of the visitors in a given location, which is however not systematic. It does not take place regularly, and in all aesthetically valuable locations, it is often associated with the use of other cultural ES, and thus, it is not possible to carry out reliable comparisons of individual sites.

5.3 Natural and Cultural Heritage: Intellectual and Scientific Values (C3)

5.3.1 Definition and Brief Characteristics of ES

The cultural ES also include the ES natural and cultural heritage. According to MEA (2005) and TEEB (2013), this service is associated with the existence of a traditional landscape formed by a specific relationship of people and nature (e.g. specific forms of dispersed settlements, vineyards, orchards, artificial water bodies and ponds, traditional agriculture, etc.). Cultural and natural heritage represents our past legacy, our current lives today and what we pass on to future generations. It includes mainly the tangible objects which were produced and used by previous generations, from small home tools to large buildings, monuments, places and landscapes. It also includes intangible elements – symbolic products of human creativity and imagination such as music, art, poetry and prose, knowledge and know-how which contribute to understanding of the heritage of the society or any partial group (UK NEA 2011).

Heritage is a broad and general term. It includes the value which an individual or society considers valuable and worth being preserved, protected, catalogued, displayed, restored and admired (Kersel and Luke 2015). The heritage is often artificially divided into natural and cultural components. Importantly, these two components are closely related, and their value overlaps in several directions.

Natural heritage defines a set of natural components and geographical structures. The value of natural heritage is important in particular in terms of biodiversity conservation, ecosystem functionality, conservation of plant and animal species and their communities and the preservation of valuable natural ecosystems. From this aspect, the most valuable ecosystems include the natural ecosystems with a rich presence of rare and endangered plant species, their communities, and animal species associated with them. Many are part of protected areas. In Slovakia, there are 9 national parks, 14 protected landscape areas and 1097 small-scale protected areas, 4 biosphere reserves, 642 special areas of conservation and 41 protected bird areas. Furthermore, the international status is represented by the RAMSAR areas and the UNESCO heritage sites.

Cultural heritage is usually defined as the heritage of biophysical functions, material and non-material attributes of a group or society which are inherited from past generations and maintained at the necessary quality for the benefit of the next generation (Czepczynski 2008). Cultural heritage forms an integral part of the historical relationships between people, society and ecosystems. The landscape originates from long and complex relationships between natural and anthropogenic factors which interact with and still modify the landscape in space and time (Reynard and Coratza 2016). Cultural landscapes represent cultural values and contribute to community identity (Stephenson 2008). Culture is not static and is often an important indicator of ecosystem and landscape change. Many elements of cultural heritage are declared cultural monuments. As of 4 November 2019, there were 9990 immovable cultural heritage monuments and 15,169 movable cultural heritage monuments (available online: http://www.pamiatky.sk/sk/page/pamiatkovy-fond) registered in the SR.

Only a summary of natural and cultural values gives each landscape the importance of a heritage which leads to community support for its conservation and improvement (Aplin 2002). Natural and cultural heritage sites are represented most commonly by national parks and protected areas, mountain areas, caves, mineral and thermal spring sites, cultural-historical conservation sites and zones, folk architecture reservations, national cultural landmarks, nature museums and the like. Cultural heritage also includes traditions, rituals, performing arts, social customs, festive events, knowledge and experiences about nature and society, which can be collectively referred to as *intangible cultural heritage* or *living heritage*. ES assessment of the intangible cultural heritage is not realized due to its nature and the lack of definite boundaries.

5.3.2 Methods Used to Assess and Identify ES

The importance of cultural ES, including natural and cultural heritage, is widely accepted, but given their *intangible* and *subjective* nature, their biophysical or monetary valuation is relatively complicated. As reported by Schröter et al. (2019), most cultural ES are difficult to measure, monitor and model. The value assigned to natural and cultural heritage services often depends on individual and cultural assessment of their contribution to well-being (Charles and Dukes 2007). Ecological and natural resources, or aspects which form an integral part of cultural heritage, are often public goods and therefore do not have a market price that reflects their value (Daniel et al. 2012; Hølleland et al. 2017). Therefore, noneconomic assessment methods are used to assess these services, in particular sociocultural methods with the use of participatory methods such as stakeholder workshops, questionnaires and personnel interviews, where the population's attitude to these valuable structures is ascertained. The perception of the significance of these structures by individual stakeholder groups is assessed, with the answers evaluated by different statistical methods. In order for the values to be effectively translated into policy-making and decision-making processes, it is important to identify also natural landscape features which are valuable from a stakeholder perspective, and multifunctional assessments need to be implemented. One of the frequently used ways of assessment of the ES natural and cultural heritage is a deliberative discussion facilitated by experts, allowing cultural and ecological values to be taken into account, as well as local and traditional stakeholder knowledge and attitude without monetary valuation (Daniel et al. 2012). For the assessment of the ES natural and cultural heritage supply, a frequently used method is the mapping of the presence of areas of visitor's interest. In particular, photo-series analysis, online map surveys and mobile phone applications are used. Often the so-called contingent valuation method is used which consists of directly assessing people's willingness to pay or accept compensation for a change in the ES in a hypothetical market (Farber et al. 2006).

Natural and cultural heritage as one of the cultural ES has also been included in several national ecosystem assessments, for example in Spain, Hungary or France. At the same time, France has chosen a specific approach in relation to natural (and cultural) heritage by separating natural heritage from the cultural ES. Natural heritage is not considered in France to be an ecosystem service with use value – on the contrary, its value is non-use (existential, altruistic). The place of service or benefit is considered one of the aspects of identity or identification (including elements of spiritual or symbolic value) between ecosystems and society. In the French assessment, alternative methods of documenting and describing the value of certain elements of ecosystems are proposed for the assessment of natural heritage. Alternatively, the assessment may also take the form of a national inventory of natural heritage features (Tibi and Therond 2017). It is rather questionable whether this analysis, description and inventory can be considered as ES assessments of natural and cultural heritage.

The assessment of the ES natural and cultural heritage is also closely linked to intellectual ES, such as science, research and education. There are many indicators for the assessment of cultural and natural heritage in the context of scientific and intellectual values, for example, the occurrence of cultural and historical monuments; number of field trips and school activities; number of seminars, workshops and conferences; occurrence of educational facilities; the number of scientific publications and studies in the territory; number of television and radio programs; number of books and information materials; and number of educational trails and panels.

Other measurable indicators include, for example, number of visitors in a given territory (per year), willingness to pay for entrance/events and admission price, the spatial extent of important areas and habitats (for birds, etc.), the occurrence of protected species and accessibility of territory. These are indicators that can be acquired and subsequently assessed in a given territory (monument, site, traditional landscape, national park, natural area).

5.3.3 The Main Types of Landscape and Ecosystems Which Provide ES

Essentially, all types of ecosystems (aquatic, forest, agroecosystem, urban ecosystem) can include and provide a natural and cultural heritage which can be further studied or explored, providing space for learning and education. Each ecosystem and each cultural landscape are in its own way unique and specific, showing a certain stage of development and thus becoming subject to ES research linked to natural and cultural heritage. Some ecosystems and landscape structures are more significant than others. Therefore, the assessment of ecosystems from this aspect is taking place at all levels, in all ecosystems and in all types of landscapes.

The types of landscapes providing this ecosystem service include territories which individuals or society consider intellectually enriching and have a footprint or legacy from the past. According to MEA (2005), this benefit is particularly provided by historically significant landscapes – cultural-historical structures of landscape. The ecosystems which provide this service are mostly related to the existence of a traditional landscape created by the specific relationship of people and nature. Based on this relationship, many specific regions have been established, such as mountain landscape in Portugal or the Alps, pastures in temperate zones of Europe, the concept of small-scale agriculture and forestry in Japan, wine regions in France, Tuscany in Italy, Napa Valley in the United States and dispersed forms of settlement in Slovakia and Romania. These landscapes represent the region as a whole and act as a trademark for a tourist offer or marketing products produced in these locations. It is the representation of the region as a whole that can be considered as a benefit of the cultural ES (Daniel et al. 2012).

5.3.4 Importance of ES in Terms of Nature and Landscape Protection in Slovakia

Cataloguing and preserving cultural and natural sites of particular importance as the common heritage of mankind is the goal of the UNESCO World Heritage Program and should be important for every society. Natural and cultural heritage includes a mixture of natural and cultural values, giving each landscape its specificity and importance, leading to the promotion from the society and the preservation of values for future generations. MEA (2005) recognizes that many societies place a high value on maintaining historically important landscapes (cultural landscapes). The preservation of cultural heritage can bring considerable synergies with the preservation of other ES, one of the motives introduced by agri-environmental programs in the European Union and the United States and the recent Satoyama Initiative in support of the United Nations Convention on Biological Diversity (Takeuchi 2010).

In Slovakia, there is quite a lot of attention devoted to the sites of natural and cultural heritage, which are legally protected (Fig. 5.10). Less attention is paid to sites and regions representing historical landscape structures with a traditional way of management, especially farming. Maintaining such forms is very demanding, and Slovakia does not have sufficient financial, technical nor human potential to maintain these forms of farming (Špulerová et al. 2017).

It is important to understand the relationship between the use of valuable land and its protection. From the point of view of protecting these values, it is necessary to respect the limits and regulations, whether set by legislation or administrators and users – visitor rules, movement in the territory and so on. The benefits provided by this kind of ES say that the different aspects of protection, use and promotion of valuable sites are all closely linked. A visitor who is interested to spiritually fuel up and observe the natural and cultural monuments is certainly not interested in the object of his interest to be devalued and otherwise destroyed. Therefore, the intention of the support of ES natural and cultural heritage is consistent with the protection of the elements which provide this service.

Fig. 5.10 The Východné Karpaty Biosphere Reserve is part of Slovakia's natural heritage, author: S. David

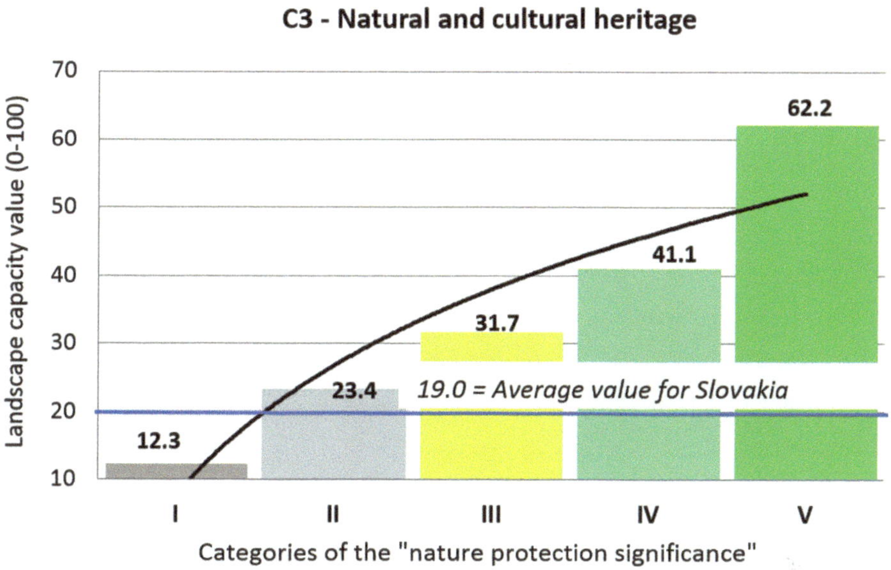

Fig. 5.11 The relationship between ecosystem service C3 and the significance of the territory of Slovakia in terms of nature and landscape protection

Figure 5.11 shows a clear positive correlation between the capacity of this ES and the significance of the territory of Slovakia in terms of nature and landscape protection. Therefore, it can be stated that the protection of nature and landscape is of key importance for the support of the ES natural and cultural heritage.

5.3.5 ES Assessment for the Territory of Slovakia

Although Slovakia is spatially small, it is naturally a very diverse landscape. In this territory, we can find a number of valuable natural and cultural sites which are being visited and directly or indirectly provide spiritual enrichment to visitors. Natural and cultural heritage has many forms of legal protection. The oldest and best-known legal instrument is the UNESCO Convention for the Protection of the World's Natural and Cultural Heritage.

In Slovakia, these are the following cultural heritage sites included in the UNESCO list: Vlkolínec Folk Architecture Reservation; Levoča, Spiš Castle and related cultural monuments (Spišská Kapitula, Spišské Podhradie, Church of the Holy Spirit in Žehra); historical town of Banská Štiavnica (Fig. 5.12) and technical monuments in its surroundings (Hodruša-Hámre, Štiavnické Bane, Banská Belá, Voznica, Vyhne, Banský Studenec, Počúvadlo, Kopanica, Kysihýbel, Antol, Ilija and 23 water reservoirs – tajch); Bardejov Town Monument Reserve and Jewish suburbium; and wooden churches in the Slovak part of the Carpathian arch

Fig. 5.12 The town of Banská Štiavnica, together with the technical monuments in the surrounding area, was added to the UNESCO List of World Cultural and Natural Heritage Sites in 1993, author: D. Štefunková

(Hervartov, Tvrdošín, Leštiny, Kežmarok, Hronsek, Bodružal, Ladomirová, Ruská Bystrá).

The second category is represented by the *natural heritage* which in Slovakia includes the following: caves in Slovenský kras and Aggtelekský kras and Carpathian beech primeval forests - cross-border territory with Ukraine (Slovak: Havešová, Rožok, Vihorlat, Stužica-Bukovské vrchy; Ukrainian: Stužica-Uzok, Čornohora, Maramoroš, Svidovec, Kuzij-Tribušany, Uhoľka-Široký Luh) (Kureková 2016).

Based on the legislation in force, other categories can be distinguished in Slovakia, where important elements of the natural and cultural landscape intersect (Hrnčiarová 2004):

- Monuments of historical vegetation (alleys, arboretums, cemeteries, city parks, parks, ornamental gardens) – declared under Act No. 49/2002 Coll. but include values of natural and cultural-historical character
- Protected areas (arboretum, botanical and other gardens, parks) and protected trees – declared under Act No. 543/2002 Coll. but also include values of a cultural-historical character

All these areas present a close link with history and the ecological, landscape and aesthetic values of the territory. They contribute to the preservation of the biological and cultural diversity of the territory, the diversity of conditions and forms of life and to the preservation of Slovakia's natural and cultural heritage.

In addition to the aforementioned elements, which are declared under the applicable laws, the cultural and historical monuments in the territory of Slovakia can also include legislative unprotected *historical landscape structures with a traditional way of management*, such as the following (Hrnčiarová 2004):

- Traditional wine-growing landscape (small-scale vineyards, terraces, stone walls, wine cellars and sheds)
- Traditional agrarian landscape (mosaic landscape structures consisting of complexes of narrowband fields, agrarian forms of relief – boundaries, terraces, walls, meadows, pastures, small woods and shrubs with scattered dwellings, haylofts, sheepfolds, sheds)
- Traditional mining landscape (adits, heaps, sinkholes, artificial water bodies – tajchs)
- Traditional landscape with various small technical structures (water mills, sawmills, forest railways, smitheries)
- Traditional fish pond landscape (small fish husbandry, fish ponds, tajchs)
- Traditional forms of settlement with original folk architecture (dispersed settlements – *lazy, kopanice, štále*)
- Traditional landscape with spa function (sanatoriums, springs, spas)

For the territory of Slovakia, the greatest attention in the area of cultural services related to natural and cultural heritage is paid to the assessment of areas with traditionally managed agricultural landscape and to the ES assessment in protected areas. The assessment of the ES in various types of traditional landscape was investigated by Špulerová et al. (2014) and Žarnovičan et al. (2018). Krnáčová et al. (2013) focused on the examples of permanent cultures of orchards and vineyards, and Petrovič (2005) focused on areas of dispersed settlements. In these model areas, among others, the importance for science and research was also assessed. This focused particularly on the ES: the use of natural systems for school excursions, research and the like.

A special example is the assessment of the ES in protected areas of Slovakia, with several studies carried out in the national parks, namely, the Vysoké Tatry NP, Slovenský raj, Veľká Fatra, Muránska planina and Malá Fatra (for a list of citations, see Chap. 1).

For the assessment of this ES for the territory of Slovakia, we have actually used several basic input layers – in addition to the landscape structure and the way of land use, these mainly included data on important natural and cultural-historical values (see Table 5.3). According to the realized assessment, the highest potential of natural and cultural heritage is concentrated in several areas – especially Vysoké Tatry, Slovenský raj, Slovenský kras, Východné Karpaty, Štiavnické vrchy and Poľana. Higher values are also achieved by the core parts of the Nízke Tatry, Malá Fatra, Veľká Fatra, Slovenské Rudohorie, Malé Karpaty and other lower mountain ranges, as well as the Podunajsko Region. As with other cultural ES, the lowest

Table 5.3 Input data for assessment of ES natural and cultural heritage

Input data/ES	C3 – Natural and cultural heritage
Capacity	Map of current landscape structure – reclassification of selected units
	Nature and landscape protection – reclassification of the significance of all types of nature conservation areas
	UNESCO World Heritage Sites
	Other significant natural resources – watercourses, natural healing resources, forest areas
	Important geological and geomorphological localities
	Land use – occurrence of specific cultural and historical landscapes (traditional use, dispersed settlements, vineyards, mining landscape)
	Specific features of the landscape – archaeological sites, historical and cultural monuments, historical parks
Demand	Data on population distribution – urban settlements and densely populated areas
	Data on the number of visitors – municipalities, regions
	Infrastructure map (location of hotels, restaurants, car parks, recreational areas and so on)
Flow	Photos on social networks – number of visitors and attractiveness of selected areas
	Statistical data on visits to selected territories – as in the case of C1

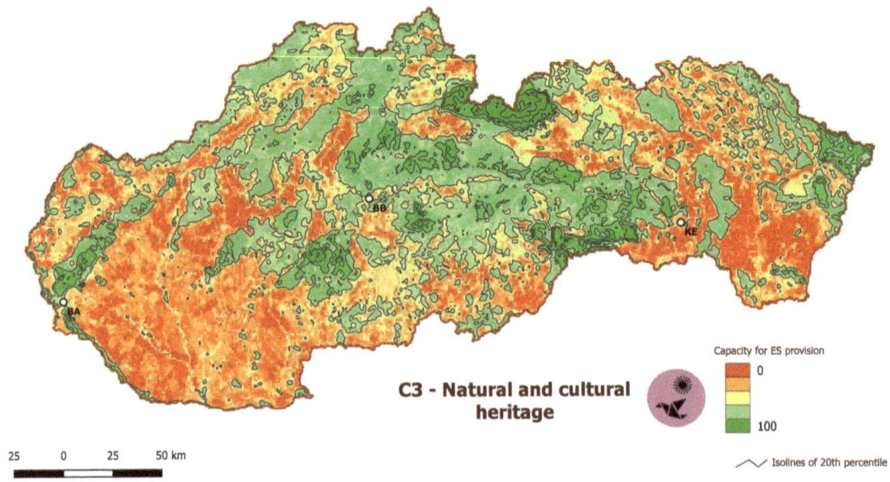

Fig. 5.13 Capacity of the landscape to provide ES natural and cultural heritage

values of the landscape's capacity are typical for the larger lowlands and basins of Slovakia (Fig. 5.13).

Slovakia has a significant potential for the provision of ES natural and cultural heritage, as there are a number of cultural and natural sites in the area. Demand for this ES is differentiated, which is largely due to the significance of the site (uniqueness, attractiveness, cultural-historical and nature-preserving value, etc.) but also due to the state of preservation and maintenance and the infrastructure built to support the site accessibility. Appropriate promotion of natural and cultural heritage also significantly supports the demand for this ES. On the other hand, the limiting factor for the use of these sites, especially natural sites, is their legislative protection.

Real use (flow) of this ES can be assessed by several indicators – similarly to previous cultural ES, it is mainly the available statistical data on number of visits to selected areas (e.g. the number of tickets sold to cultural and historical monuments, the number of visitors to UNESCO sites, the number of overnight stays). Use of the ES natural and cultural heritage can also be assessed on the basis of indirect indicators, such as the area of traditional forms of farming, which, in addition to this service, are also used to provide production services, and the area of protected areas also used for research and education. An additional indicator can also be in the form of the number of photos on social networks.

References

Aplin, G. (2002). *Heritage: Identification, conservation, and management*. Melbourne: Oxford University Press.

Barčáková, I. (2001). Prístupy k hodnoteniu estetickej (vizuálnej) kvality krajiny. *Geografický časopis, 53*, 343–356.

Beaumont, J. N., Austen, M. C., Atkins, J. P., Burdon, D., Degraer, S., Dentinho, T. P., DErous, S., Holm, P., Horton, T., Van Ierland, E., Marboe, A. H., Starkey, D. J., Townsend, M., & Zarzycki, T. (2007). Identification, definition and quantification of goods and services provided by marine biodiversity: Implications for the ecosystem approach. *Marine Pollution Bulletin, 54*(3), 253–265. https://doi.org/10.1016/j.marpolbul.2006.12.003.

Becerra-Jurado, G., Philipsen, C., & Kleeschulte, S. (2016). *Mapping and assessing ecosystems and their Services in Luxembourg – Assessment results*. Luxembourg: Le Gouvernment du Grand -Duché de Luxemburg. https://doi.org/10.13140/rg.2.1.4924.5841.

Boyd, J., & Banzhaf, S. (2007). What are ecosystem services? The need for standardized environmental accounting units. *Ecological Economics, 63*(2–3), 616–626. https://doi.org/10.1016/j.ecolecon.2007.01.002.

Brown, T. (1994). Conceptualizing smoothness and density as landscape elements in visual resource management. *Landscape and Urban Planning, 30*(1–2), 49–58. https://doi.org/10.1016/0169-2046(94)90066-3.

Burkhard, B., Kandziora, M., Hou, Y., & Müller, F. (2014). Ecosystem service potentials, flows and demands-concepts for spatial localisation, indication and quantification. *Landscape Online, 34*, 1–32.

Charles, L. H., & Dukes, S. J. (2007). Impacts of invasive species on ecosystem services. In V. Nentwig (Ed.), *Biological invasions* (pp. 217–237). Springer.

Czepczynski, M. (2008). *Cultural landscapes of post-socialist cities: Representation of powers and needs*. Aldershot-Burlington: Ashgate.

Daniel, T. C. (2001). Whither scenic beauty? Visual landscape quality assessment in the 21st century. *Landscape and Urban Planning, 54*(1–4), 267–281. https://doi.org/10.1016/S0169-2046(01)00141-4.

Daniel, T. C., Muhar, A., Arnberger, A., Aznar, O., Boyd, J. W., Chan, K. M. A., Costanza, R., Elmqvist, T., Flint, C. G., Gobster, P. H., Gret-regamey, A., Lave, R., Muhar, S., Penker, M., Ribe, R. G., Schauppenlehner, T., Sikor, T., Soloviy, I., Spierenburg, M., Taczanowska, K., Tam, J., & Von Der Dunk, A. (2012). Contributions of cultural services to the ecosystem ser-

vices agenda. *Proceedings of the National Academy of Sciences, 109*(23), 8812–8819. https://doi.org/10.1073/pnas.1114773109.
Drdoš, J. (1995). Krajinný obraz a jeho hodnotenie. *Životné prostredie, 29*(4), 202–205.
Farber, S., Costanza, R., Childers, D. L., Erickson, J., Gross, K., Grove, M., Hopkinson, C. S., Kahn, J., Pincetl, S., Troy, A., Warren, P., & Wilson, M. (2006). Linking ecology and economics for ecosystem management. *BioScience, 56*(2), 121–133. https://doi.org/10.1641/0006-3568(2006)056[0121:LEAEFE]2.0.CO;2.
Fisher, B., Turner, R. K., & Morling, P. (2009). Defining and classifying ecosystem services for decision making. *Ecological Economics, 68*(3), 643–653. https://doi.org/10.1016/j.ecolecon.2008.09.014.
Frélichová, J., Vačkář, D., Pártl, A., Loučková, B., Harmáčková, Z. V., & Lorencová, E. (2014). Integrated assessment of ecosystem services in the Czech Republic. *Ecosystem Services, 8*, 110–117. https://doi.org/10.1016/j.ecoser.2014.03.001.
Fűzyová, Ľ., Lániková, D., & Novorolský, M. (2009). Economic valuation of Tatras national park and regional environmental policy. *Polish Journal of Environmental Studies, 18*(5), 811–818.
Gallagher, W. (1993). *The power of place: How our surroundings shape our thoughts, emotions, and actions*. New York: Poseidon Press.
Getzner, M. (2009). *Economic and cultural values related to protected areas. Part A: Valuation of ecosystem services in Tatra (PL) and Slovensky Raj (SK) national parks*. Final report. http://www.minzp.sk/files/sekcia-ochranyprirodyakrajiny/ps-543/final-report-part-b-pl-sk_getzner_201003.pdf. Accessed 04 Apr 2019.
Giglio, S., Bertacchini, F., Bilotta, E., & Pantano, P. (2019). Using social media to identify tourism attractiveness in six Italian cities. *Tourism Management, 72*, 306–312. https://doi.org/10.1016/j.tourman.2018.12.007.
Hanley, N., Wright, R. E., & Koop, G. (2002). Modelling recreation demand using choice experiments: Climbing in Scotland. *Environmental and Resource Economics, 22*(3), 449–466. https://doi.org/10.1023/A:1016077425039.
Hølleland, H., Skrede, J., & Holmgaard, B. S. (2017). Cultural heritage and ecosystem services: A literature review. *Conservation and Management of Archaeological Sites, 19*(3), 210–237. https://doi.org/10.1080/13505033.2017.1342069.
Hrnčiarová, T. (2004). Prírodné a kultúrne aspekty krajiny a jej potenciál. *Životné Prostredie, 38*(2), 61–65.
Hrnčiarová, T., Kenderessy, P., Špulerová, J., Vlachovičová, M., Piscová, V., & Dobrovodská, M. (2018). Status and outlook of hiking trails in the central part of the low Tatra Mountains in Slovakia between 1980–1981 and 2013–2014. *Journal of Mountain Science, 15*, 1615–1632. https://doi.org/10.1007/s11629-017-4690-3.
Jakab, I., & Petluš, P. (2013). The use of viewshed analysis in creation of maps of potential visual exposure. In *GIS Ostrava 2013 – Geoinformatics for city transformation*. Ostrava: VSB Technical University Ostrava, Faculty of mining and geology, Institute of Geoinformatics. http://gisak.vsb.cz/GIS_Ostrava/GIS_Ova_2013/proceedings/papers/gis2013508aaf204a6ff.pdf. Accessed 6 Apr 2019.
Jančura, P., Bohálová, I., Slámová, M., & Mišíková, P. (2010). Method of identification and assessment of characteristic landscape appearance. *Bulletin of the Ministry of Environment of Slovak Republic, 18*(Part 1b), 2–51.
Jurko, A. (1990). *Ekologické a socio-ekonomické hodnotenie vegetácie*. Bratislava: Píroda.
Kemp, D. R., & Michalk, D. L. (2007). Towards sustainable grassland and livestock management. *The Journal of Agricultural Science, 145*(6), 543–564. https://doi.org/10.1017/S0021859607007253.
Kersel, M. M., & Luke, C. H. (2015). Civil societies? Heritage diplomacy and neo-imperialism. In L. Meskell (Ed.), *Global heritage: A reader* (p. 2015). Chichester: Wiley.
Koschke, L., Fuerst, C., Frank, S., & Makeschin, F. (2012). A multi-criteria approach for an integrated land-cover-based assessment of ecosystem services provision to support landscape planning. *Ecological Indicators, 21*, 54–66. https://doi.org/10.1016/j.ecolind.2011.12.010.

Krause, C. (1991). Die Praxis der Landschaftsbilderfassung am Beispiel StrassenBau, BFANL (Hrsg.). *Landschaftsbild-Eingriff-Ausgleich*, pp. 121–141.

Krnáčová, Z., Barančok, P., Pavlíčková, K., & Platková, A. (2013). *The methodological approaches comparison of the landscape offer and ecosystems benefits evaluation for ecological tourism (ECOTOUR) development options (Bratislava IV example – cadastre Devín).* In: Abstracts of RegioResources 21, University of Catania, Catania.

Kureková, I. (2016). Prírodné a kultúrne dedičstvo (nielen) Slovenska. *Enviromagazín, 5*, 12–13.

Lankia, T., Kopperoinen, L., Pouta, E., & Neuvonen, M. (2015). Valuing recreational ecosystem service flow in Finland. *Journal of Outdoor Recreation and Tourism, 10*, 14–28. https://doi.org/10.1016/j.jort.2015.04.0.

Lieskovský, J., Rusňák, T., Klimantová, A., Izsóff, M., & Gašparovičová, P. (2017). Appreciation of landscape aesthetic values in Slovakia assessed by social media photographs. *Open Geosciences, 9*(1). https://doi.org/10.1515/geo-2017-0044.

Lotan, A., Kost, R., Mandelik, Y., Peled, Y., Chakuki, D., Shamir, S. Z., & Ram, Y. (2018). National scale mapping of ecosystem services in Israel – Genetic resources, pollination and cultural services. *One Ecosystem, 3*, e25494. https://doi.org/10.3897/oneeco.3.e25494.

MEA – Millennium Ecosystem Assessment. (2005). *Ecosystems and human well-being: Wetlands and water synthesis*. Washington, DC: World Resources Institute.

Milcu, A. I., Hanspach, J., Abson, D., & Fischer, J. (2013). Cultural ecosystem services: A literature review and prospects for future research. *Ecology and Society, 18*(3), 44. https://doi.org/10.5751/ES-05790-180344.

Mišíková, P. (2002). *Vnímanie krajiny (príkladová štúdia v lokalite Sĺňava – Piešťany).* Papet presented at the Nové trendy v krajinnej ekológii, Comenius University in Bratislava, Piešťany, 24–26 October 2002.

Nahuelhual, L., Carmona, A., Lozada, P., Jaramillo, A., & Aguayo, M. (2013). Mapping recreation and ecotourism as a cultural ecosystem service: An application at the local level in Southern Chile. *Applied Geography, 40*, 71–82. https://doi.org/10.1016/j.apgeog.2012.12.004.

Nohl, W. (1991). Konzeptionelle und methodische Hinweise auf landschaftsästhetische Bewertungskriterien für die Eingriffsbestimmung und die Festlegung des Ausgleichs. *Landschaftsbild-Eingriff-Ausgleich*, 59–75.

Oťaheľ, J. (1999). Visual landscape perception: Landscape pattern and aesthetic assessment. *Ekologia Bratislava, 18*(1), 63–74. https://www.researchgate.net/publication/299069946_Visual_landscape_perception_Landscape_pattern_and_aesthetic_assessment. Accessed 6 Apr 2019.

Oťaheľ, J. (2003). Visual quality of the landscape: Approaches to analysis. *Ekologia Bratislava, 22*(2), 150–160.

Othman, J. (2015). Assessing scenic beauty of nature-based landscapes of Fraser's Hill. *Procedia Environmental Sciences, 30*, 115–120. https://doi.org/10.1016/j.proenv.2015.10.020.

OZ Pronatur. (2014). *Hodnotenie ekosystémových služieb v NP Muránska planina.* https://www.minzp.sk/files/oblasti/ochrana-prirody-a-krajiny/ekosystemove-sluzby/povazan_mp_pronatur.pdf. Accessed 6 Apr 2019.

Palkechová, L., & Kozáková, J. (2015). Ekonomika agroturistických prevádzok vo vybraných agrosubjektoch SR. *Ekonomika poľnohospodárstva, 15*, 69–85.

Pásková, M. (2008). In Gaudeamus Univerzita Hradec Králové (Ed.), *Udržitelnost rozvoje cestovního ruchu* (p. 298).

Petrovič, F. (2005). *Vývoj krajiny v oblasti štálového osídlenia Pohronského Inovca a Tribeča.* Bratislava: ÚKE SAV.

Piscová, V., Hrnčiarová, T., Hreško, J., Dobrovodská, M., Izakovičová, Z., Iszoff, M., Kalivoda, H., Kanka, R., Kenderessy, P., Považan, R., Roháč, J., Slabejová, M., Slivinský, J., Špulerová, J., Švajda, J., Tepercer, J., Vlachovičová, M., Brorvský, I., & Hurta, V. (2018). *Využívanie vysokohorskej krajiny a jeho dôsledky na zmenu prostredia (na príklade Tatier a Nízkych Tatier).* Bratislava: VEDA.

Pourová, M. (2000). *Agroturistika, možnosti rozvoje a perspektiva v České republice*. Praha: KUFR.

Považan, R. (2013). Rekreačné hodnoty NP Veľká Fatra. *Acta Universitatis Matthiae Belii séria envirnomentálne manažérstvo, 15*(1), 82–94.

Považan, R., Getzner, M., & Švajda, J. (2014a). Value of ecosystem services in mountain national park. Case study of Veľká Fatra National Park (Slovakia). *In: Polish Journal of Environmental Studies, 23*(5), 1699–1710. ISSN 1230-1485.

Považan, R., Getzner, M., & Kadlečík, J. (2014b). Hodnotenie ekosystémových služieb v chránených územiach Karpát so zameraním na Slovensko – Príručka pre rýchle hodnotenie. *Quaestiones rerum naturalium, 1*(2), 7–44.

Považan, R., Getzner, M., & Švajda, J. (2015). On the valuation of ecosystem services in Muránska Planina National Park (Slovakia). *Eco. Mont. Journal on Protected Mountain Areas Research and Management, 7*(2), 61–69. https://doi.org/10.1553/eco.mont-7-2s61.

Reynard, E., & Coratza, P. (2016). The importance of mountain geomorphosites for environmental education. Examples from the Italian Dolomites and the Swiss Alps. *Acta geographica Slovenica, 56*(2), 91–303. https://doi.org/10.3986/AGS.1684.

Santos, T., Nogueira, M., & Vasco, A. (2016). Recreational activities in urban parks: Spatial interactions among users. *Journal Outdoor recreation and Tourism, 15*, 1–9. https://doi.org/10.1016/j.jort.2016.06.001.

Schmidt, K., Walz, A., Jones, I., & Metzger, M. J. (2016). The sociocultural value of upland regions in the vicinity of cities in comparison with urban green spaces. *Mountain Research and Development, 36*(4), 465–474. https://doi.org/10.1659/MRD-JOURNAL-D-16-00044.1.

Schirpke, U., Scolozzi, R., Da re, R., Masiero, M., Pellegrino, D., & Marino, D. (2018). Recreational ecosystem services in protected areas: A survey of visitors to Natura 2000 sites in Italy. *Journal of Outdoor Recreation and Tourism, 21*, 39–50. ISSN 2213-0780.

Schröter, M., Bonn, A., Klotz, S., Seppelt, R., & Baessler, C. (2019). *Atlas of ecosystem services. Drivers, risks, and societal responses*. Springer.

Sottini, V. A., Bernetti, I., Pecchi, M., & Cipollaro, M. (2018). Visual perception of the rural landscape: A study case in Val di Chiana aretina, Tuscany (Italy). *Aestimum, 72*, 5–26. https://doi.org/10.13128/Aestimum-23967.

Špulerová, J., Dobrovodská, M., Štefunková, D., Bača, A., & Lieskovský, J. (2014). *Biodiversity of traditional agricultural landscapes in Slovakia and their threats. Biocultural Landscapes*. Dordrecht: Springer.

Špulerová, J., Štefunková, D., Dobrovodská, M., Izakovičová, Z., Kenderessy, P., Vlachovičová, M., Lieskovský, J., Piscová, V., Petrovič, F., Kanka, R., Bača, A., Barančoková, M., Bezák, P., Bezáková, M., Boltižiar, M., Mojses, M., Dubcová, M., Gajdoš, P., Gerhátová, K., Izsóff, M., Kalivoda, H., Miklósová, V., Drábová, M., Šatalová, B., Krištín, A., Dankaninová, L., Kalivodová, E., Majzlan, O., Mihál, I., Stašiov, S., Šolomeková, T., Ambros, M., Baláž, I., & Halabuk, A. (2017). Historické štruktúry poľnohospodárskej krajiny Slovenska. Bratislava: Veda. 144 s. ISBN 978-80-224-1570-5.

Štefunková, D. (2004). Príklad hodnotenia vizuálnej kvality krajiny v prostredí geografických informačných systémov na vybranom modelovom území. In V. Herber (Ed.), *Fyzicko-geografický sborník 2, Kulturní krajina*. Brno: Masaryk University in Brno.

Štefunková, D., & Cebecauer, T. (2006). Visibility analysis as a part of landscape visual quality assessment. *Ekológia (Bratislava), 25*(1), 229–239.

Stephenson, J. (2008). The cultural values model: An integrated approach to values in landscapes. *Landscape Urban Planning, 84*(2), 127–139. https://doi.org/10.1016/j.landurbplan.2007.07.003.

Supuka, J., & Vreštiak, P. (1984). *Základy tvorby parkových lesov (a iných rekreačne využívaných lesov)*. Bratislava: Veda.

Švajda, J. (2009). Contribution for improvement of visitor monitoring in the Tatra National Park. *Eco. Mont-Journal on Protected Mountain Areas Research, 1*, 13–18. https://doi.org/10.1553/ecomont2s13.

Švajda, J., & Fenichel, E. P. (2011). Evaluation of integrated protected area Management in Slovakian National Parks. *Polish Journal of Environmental Studies, 20*(4), 1053–1060. http://www.pjoes.com/Evaluation-of-Integrated-Protected-Area-r-nManagement-in-Slovakian-National-Parks,88651,0,2.html. Accessed 10 Apr 2019.

Švajda, J., Masny, M., Korony, S., Mezei, A., Machar, I., & Taczanowska, K. (2018). Visitor profiling using characteristics of socio-demographic and spatial behavior as tools to support the management of protected mountain areas. *Geografie, 123*(4), 461–478. https://doi.org/10.37040/geografie2018123040461.

Takeuchi, K. (2010). Rebuilding the relationship between people and nature: The Satoyama initiative. *Ecological Research, 25*, 891–897. https://doi.org/10.1007/s11284-010-0745-8.

TEEB – The Economics of Ecosystems and Biodiversity. (2013). *Guidance manual for TEEB country studies*. Version 1.0.

Tibi, A., & Therond, O. (2017). *Evaluation des services écosystémiques rendus par les écosystèmes agricoles*. Une contribution au programme EFESE. Synthèse du rapport d'étude. France. INRA Science and Impact. https://www.inrae.fr/sites/default/files/pdf/efese-services-ecosystemiques-rendus-par-les-ecosystemes-agricoles-synthese-2.pdf

UK NEA. (2011). *The UK National Ecosystem Assessment: Synthesis of the key findings*. Cambridge: UNEP-WCMC.

Vačkář, D., Frélichová, J., Lorencová, E., Pártl, A., Harmáčková, Z., & Loučková, B. (2014). *Metodologický rámec integrovaného hodnocení ekosystémových služeb v České republice*. http://www.ecosystemservices.cz/cs/metodologicky-ramec-integrovaneho-hodnoceni-ekosystemovych-sluzeb-v-ceske-republice. Accessed 13 Apr 2019.

Vandewalle, M., Sykes, M. T., Harrison, P. A., Luck, G. W., Berry, P., Bugter, R., Dawson, T. P., Feld, C. K., Harrington, R., Haslett, J. R., Hering, D., Jones, K. B., Jongamn, R., & Lavorel, S. (2009). *Review paper on concepts of dynamic ecosystems and their services. The Rubicode Project Rationalising Biodiversity Conservation in Dynamic Ecosystems*. http://www.rubicode.net/rubicode/RUBICODE_Review_on_Ecosystem_Services.pdf. Accessed 10 Apr 2019.

Vološčuk, I., Sabo, P., Škodová, M., Švajda, J., & Lepeška, T. (2016). *Dynamika krajinnej štruktúry a diverzita ekosystémov Krivánskej Fatry*. Banská Bystrica: Matej Bel University.

Wöbse, H. H. (1991). Landschaftsästhetik und ihre Operationalisierungsmoglichkeiten bei der Anwendung des § 8 Bundesnaturschutzgesetz. *Landschaftsbild-Eingriff-Ausgleich*, 31–35.

Žarnovičan, H., Pavličková, K., & Kollár, J. (2018). Ovocné sady ako fenomén kultúrnej krajiny novobanskej štálovej oblasti. *Životné prostredie, 52*(4), 247–251.

Zube, E. H., Sell, J. L., & Taylor, J. G. (1982). Landscape perception: Research, application and theory. *Landscape Planning, 9*(1), 1–33. https://doi.org/10.1016/0304-3924(82)90009-0.

Part III
Synthesis

Chapter 6
Synthesis of Ecosystem Services Assessment in Slovakia

Peter Mederly, Ján Černecký, Jana Špulerová, Zita Izakovičová, František Petrovič, and Viktória Ďuricová

Abstract This chapter provides comprehensive findings for the three basic groups of ES. Five provisioning ES are essential for Slovakia – agricultural crops; timber and fibre; drinking water; freshwater and fish; and game and wildfood. Regulatory ES represent the regulation of natural processes – erosion and natural hazards mitigation; runoff and flood protection; local and global climate regulation; and air and water quality regulation. Supporting ES enable the appropriate course of natural functions and processes – as biodiversity promotion; pollination; pest and disease control; or soil formation. Cultural ES are the intangible benefits of nature for people, such as recreation and tourism; landscape aesthetics; and natural and cultural heritage. As a synthesis, the overall landscape capacity for ES provision is expressed, as an average of the main ES groups. Regarding landscape types, the high value of the ES capacity is documented for mountains and sub-mountain areas, while the low capacity is typical for lowlands and open basin areas. Also, the relationships between land use and ES are evaluated. Results confirm the generally accepted fact that forest ecosystems are the most important for the ES provision (mainly decidu-

P. Mederly (✉) · F. Petrovič
Faculty of Natural Sciences, Constantine the Philosopher University in Nitra, Nitra, Slovakia
e-mail: pmederly@ukf.sk; fpetrovic@ukf.sk

J. Černecký
Faculty of Natural Sciences, Constantine the Philosopher University in Nitra, Nitra, Slovakia

State Nature Conservancy of the Slovak Republic, Banská Bystrica, Slovakia

Institute of Landscape Ecology of the Slovak Academy of Sciences, Bratislava, Slovakia
e-mail: jan.cernecky@sopsr.sk

J. Špulerová · Z. Izakovičová
Institute of Landscape Ecology of the Slovak Academy of Sciences, Bratislava, Slovakia
e-mail: jana.spulerova@savba.sk; zita.izakovicova@savba.sk

V. Ďuricová
State Nature Conservancy of the Slovak Republic, Banská Bystrica, Slovakia

Faculty of Natural Sciences, Matej Bel University, Banská Bystrica, Slovakia
e-mail: viktoria.duricova@sopsr.sk

P. Mederly, J. Černecký (eds.), *A Catalogue of Ecosystem Services in Slovakia*,
https://doi.org/10.1007/978-3-030-46508-7_6

ous forests), and urbanized areas (mainly industrial and technical infrastructure) are the least important. Finally, the crucial importance of nature and landscape protection was confirmed – not only for the healthy state of ecosystems but also for the fulfilment of their functions and services.

6.1 Provisioning Ecosystems Services

Provisioning ES are one of the main ES groups, most likely perceived and directly appreciated by most people. On the basis of the different classification systems (see Table 1.1 in Chap. 1), this includes material products and goods from ecosystems, providing nutrition, materials and energy, especially *biomass for food, drinking water and water for other purposes* and *biomass for use, abiotic materials and substances* and *energy sources*. For the pilot ES assessment, based on the opinion of MAES process experts representing the different ES assessment institutions in Slovakia, we selected five ES, including agricultural and forest biomass, drinking water and freshwater and complimentary food sources from different types of ecosystems.

P1 biomass – Agricultural crops are mainly based on the production capacity of soils and climatic-hydrological conditions. The spatial distribution of the landscape's capacity to provide this ES is therefore significantly different from that of most other provisioning ES. This ES is actually used in the agricultural production process; it is one of the *most visible*, and in terms of assessment of ES, it is the best developed. The problem is that with the intensive use of this ES, the use most of all the other ES is largely suppressed (even excluded). Especially agriculture and its practices directly affecting more than half of Slovakia's territory are extremely important not only for the use of this ES but also for the possibility of maintaining and providing other production- and most non-production ES.

P2 biomass – Timber and fiber are sometimes simplified as a complement to the previous ES because it is actually used mainly in the form of forestry. However, this is not so clear because agricultural ecosystems and other types of landscape are also involved to some extent in the provision of this ES. However, it is clear that forestry is the main factor in using and restoring this potential. Unlike agriculture with an annual and seasonal utility cycle, wood biomass benefits are mostly associated with decades-long periods – and this is a *major problem in using this ES*. Woody plants as its carrier also play a key role in providing other provisioning and, in particular, regulatory and supporting ES. *A one-time benefit from this ES* (most often through the logging of forests or small woods) can cause a *loss of benefit* in terms of the amount of other ES for decades. This is a fact which is completely neglected in sectoral landscape management in order to maximize immediate benefit.

P3 drinking water and *P4 freshwater* are closely related ES which are sometimes understood and assessed as one entity. Drinking water is crucial for the survival of humans and animals; freshwater is particularly important in terms of human economic activities, living conditions excluded and the overall condition of ecosystems. The capacity of the landscape to provide these ES depends mainly on the abiotic conditions and processes (in particular rainfall-runoff properties, precipitation balance, hydrogeological properties); ecosystem status and the overall quality of the environment are also important, especially for drinking water. The spatial projection of the landscape potential for these ES is different from the other ES, which is caused by the above facts. The landscape capacity for drinking water is concentrated in larger units with the protection of surface and underground resources; the capacity for freshwater is associated mainly with hydrogeological units with a positive balance of rainfall-runoff regime. Wider river valleys and floodplains with accumulation of quaternary gravels are of particular importance (Žitný ostrov area is of European significance in this respect). It should also be mentioned that other functions and services are sometimes restricted by the use of this ES – especially in the case of building hydropower, large water reservoirs, but also in excess of water abstraction.

P5 fish and game/wildfood depend mainly on the predominant land use, quality of the environment and, in the case of game animals, also on the regulatory intervention of humans. To a large extent, it is linked to ES P2 and dominates in lower and medium-altitude mountain ranges, but lowland and basin areas also have some potential, especially their submountain parts and areas near to larger watercourses and water bodies. The use of this ES does not fundamentally affect the benefits of other ES – it is less conflicting in this respect.

Various methods are used for the assessment of provisioning ES, including mainly the biophysical and economic ones. Capacity is expressed, e.g. with modelling of related ecosystem functions, processes and production capability, with the common use of spatial GIS models. Real use and demand for ES are also expressed through monetary methods, as provisioning ES are mostly part of the markets.

With regard to the overall spatial projection of the capacity of the landscape of Slovakia to provide provisioning ES Fig. 6.1), *the highest values are achieved by small discontinuous areas within some mountain ranges* (especially Strážovské vrchy, Veľká Fatra; partly Nízke Tatry, Malé Karpaty, Považský Inovec, Slovenský

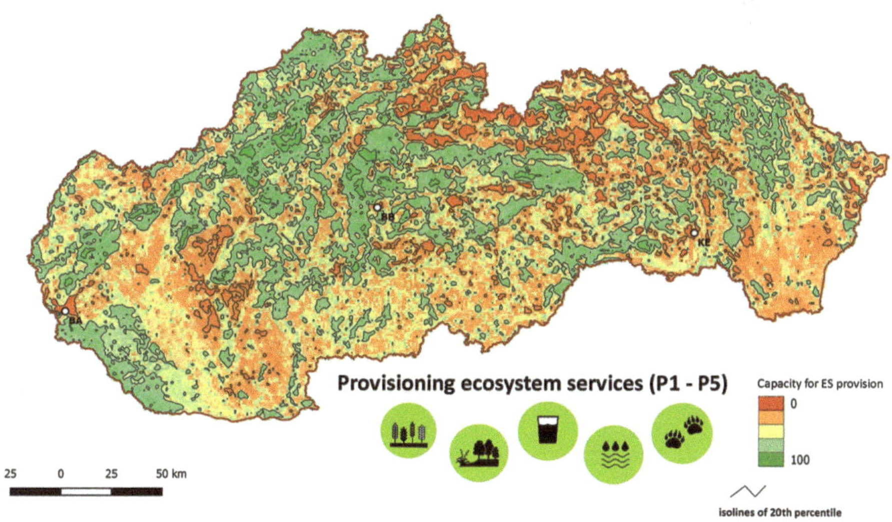

Fig. 6.1 The total capacity of the landscape to provide provisioning ES

kras). In addition to most of the lower and middle mountain ranges, high values are also achieved by sub-mountain areas and some parts of lowlands and basins. The specific area is the Podunajsko and the Žitný ostrov, which also have a high potential in terms of provisioning ES, given by their importance in terms of providing drinking and freshwater. The landscape's lowest capacity to provide provisioning ES is typical for urbanized and densely populated areas, as well as for lesser-productive and non-forested parts of lowlands and higher river basins. The Tatry region and the highest parts of the other high mountain ranges have a specific position with very little capacity for provisioning ES.

By using provisioning ES, there occurs an *abstraction* of matter and energy from the ecosystems providing the given ES. Therefore, it is very important to know their recovery capacity in terms of the time of recovery of the necessary production functions of ecosystems. Some ES are used practically constantly and have a continuous recovery ability (e.g. water), others are seasonal (agricultural and forest crops), and the timber biomass has a significantly longer recovery cycle. Another related issue is that while the use of some provisioning ES is not in principle threatening to other ES (partly water, game, wild berries), the use of agricultural crops and forest biomass largely limits the possibilities of using other ES – thus causing the so-called trade-offs (limits, conflicts of interests) not only from the point of view of some provisioning but also most of the regulatory and cultural ES.

The landscape capacity for provisioning ES as a whole compared to other groups of ES is least related to biodiversity, nature and landscape protection. The variance between the average value for the whole territory of Slovakia and the values achieved for the five basic categories significance of the territory of Slovakia in terms of nature and landscape protection is the smallest of all ES groups (Fig. 6.2). It can be

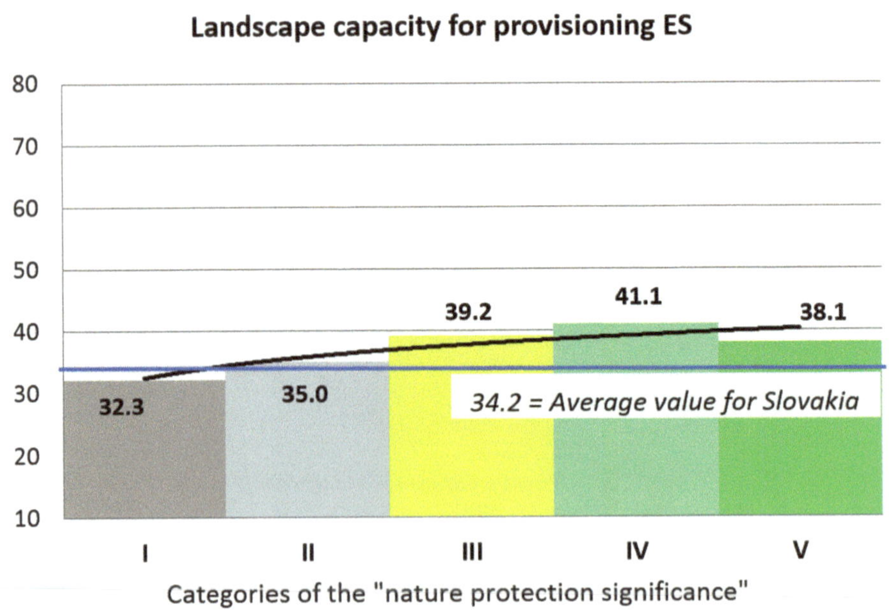

Fig. 6.2 The relationship between provisioning ecosystem services and the significance of the territory of Slovakia in terms of nature and landscape protection

said that between categories III. and V., the degree of significance of the territory, there is virtually no difference in terms of capacity to provide provisioning ES. The lowest capacity is associated with a territory which is the least important in terms of nature and landscape protection, which is due to the fact that most provisioning ES is related to ecosystems with a higher proportion of forests and natural vegetation (except for P1 biomass production). However, even here the value is only insignificantly lower (by 2%) compared to the national average, which shows the relative balance within the territory.

6.2 Regulatory Ecosystems Services and Supporting Ecosystems Functions

Regulatory ES represent the benefits of regulating processes in ecosystems, especially their abiotic components; at the same time, they contribute to improving air quality (R1), water quality (R2), and also to regulation of threatening processes such as erosion and other natural hazards (R3), floods (R4) or climate risks and extremes (R5). These ES are all closely related. Deposition of pollutants from the atmosphere in the soil and vegetation can significantly reduce their concentration in other environmental components (water, air) and thus reduce the adverse effects on human health and contribute to the provision of other ES (e.g. cultural, such as recreation, provisioning – provision of drinking/freshwater). The range of regulatory effects in pollutant deposition depends on many environmental factors, e.g. in case of air quality regulation, also from air turbulence, habitat type and duration of foliage.

Regulatory ES have a significant impact on the regulation of natural processes – erosion and natural hazards, runoff and flood protection as well as climate regulation. These landscape processes are related to land use, geological bedrock properties and slope inclination, the rainfall-runoff regime in basins as well as the protective effect of vegetation. The spatial structure of the vegetation and its properties plays an important role in soil protection and slope stabilization, water retention and climate regulation.

Terrestrial ecosystems play an important role in regulatory ES. The main media facilitating the proper functioning of water and air quality regulation include vegetation, soil and soil biota and wetland ecosystems (the metabolic activity of plants and microorganisms). Ecosystems contribute to improving the quality of individual environmental components (air, water, soil). Forests and other wooded areas are the most important ecosystems for air quality regulation, climate change and erosion control and other processes. A very good anti-erosion effect of vegetation is provided also by permanent grassland areas. For regulation of the effects of slope processes, the wooded parts of hills, highlands and mountainous areas are the most important. Riparian and non-forest vegetation are also important in the regulation of runoff conditions. It is the spatial extent and quality of urban vegetation that is important for climate regulation, as there is the greatest demand for this ES in urban areas.

Biophysical methods (or combined with economic methods) are used in particular to assess these ES. Suitable indicators for air quality regulation include atmospheric gas flow, atmosphere/air purification capacity and pollutant content/level in the atmosphere, dry deposition rate (potential), air pollutant removal (real production) and human exposure (demand). For the assessment of water quality regulation, the indicators include land use, hydrogeological properties, soil quality as well as vegetation properties – its spatial structure (coverage, biomass volume), naturalness, diversity and nutrient cycle.

A wide range of models is being used to assess the ES erosion and natural hazard regulation, runoff mitigation and flood risks regulation. Used indicators include land use, relief, the occurrence of real processes (landslides and erosion), soil parameters (depth, texture, retention capacity), state of aquatic ecosystems as well as the vegetation properties (its distribution, coverage and spatial structure). Modelling tools are also used to assess global climate regulation. The issue of erosion and other slope processes as well as the modelling of flood risk is very well developed and known for the territory of Slovakia, unlike the assessment of other ES. When considering the economic methods, it is possible to use, for example, the contingent valuation methods, cost savings or replacement costs for the air quality regulation or climate regulation.

A separate group of the ES is formed by the so-called supporting (ecological) functions and services. The most important ES include the following: (R7) biodiversity promotion; (R8) lifecycle maintenance/pollination; (R9) pest and disease control; and (R10) maintenance of soil formation and composition. However, there are many other ES which are important, e.g. decomposition function to maintain ecological stability and other services which are ignored in most assessments.

As is the case with typical regulatory ES, nature and nature-based habitats have the greatest capacity to provide supporting ES, due to their functions and ability to participate in ecological processes such as primary production, photosynthesis, reproductive capacity, pollination, nutrient cycle, soil formation and fertility maintenance. Plants, animals but also invisible fungi and microbes, which form a network of interconnections, structures and functions and are also influenced by the abiotic environment, contribute to the provision of ES. Also important are the soils which perform a number of basic functions in the landscape, such as nutrient cycle, water regulation, habitat and biodiversity protection, filtering and buffering as well as habitat stability itself. That is why the relationship between the landscape's capacity to provide these ES and the significance of the territory of Slovakia in terms of nature and landscape protection is clearly positive (Fig. 6.3).

Supporting ES have a significant impact on the provision of other ES as well as on the provision of natural functions and processes, so in some classifications (e.g. MEA 2005), they are referred to as a separate group. The potential of providing these ES is largely dependent on the ecosystem types, their status and the land use in the immediate and distant surroundings. Fertile and healthy soil is needed for habitat sustainability and food production. A landscape with a high proportion of habitats in a favourable state is much more stable, but if one wants to use some of the benefits of the landscape to achieve ecological stability, it is necessary to look

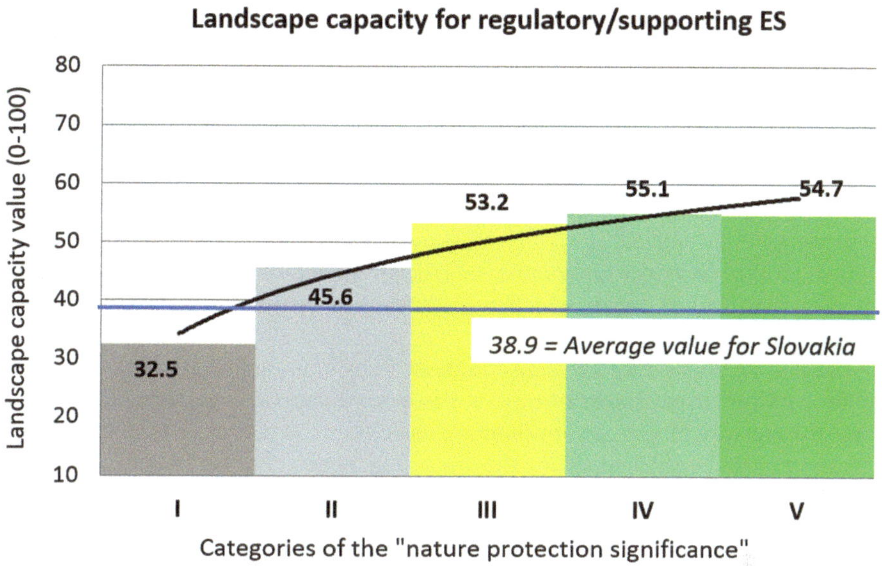

Fig. 6.3 The relationship between regulatory and supporting ecosystem services and the significance of the territory of Slovakia in terms of nature and landscape protection

for and find a carrying capacity for various anthropogenic activities so as to avoid irreversible disturbance of their recovery rate.

In the case of supporting ES, it is sometimes difficult to distinguish between the potential and the actual flows/contributions of the ecosystems to the provision of the given ES. Therefore selected biophysical service flow indicators are most often used to assess their capacity, by which they contribute to preventing or minimizing any damages if the flows of services and processes in ecosystems were disrupted. The correlation between these ES has also been reflected in a number of joint indicators for their assessment.

Biodiversity promotion is focused mainly on the conservation of biodiversity, its favourable habitat status as well as the protection of rare and endangered species, and these characteristics are most commonly used as indicators for assessing this ES.

Lifecycle maintenance/pollination is essential for maintaining and promoting the biodiversity of most wild plants, as well as for the fertility and stability of crop production dependent on pollination. For this ES, the key indicators include the habitat possibilities for nesting of pollinators as well as flower sources (the type of ecosystems and their species composition).

Natural and seminatural as well as anthropogenic ecosystems are characterized by the ability to provide *pests and diseases control* through genetic variations of plants and animals. The performance of this service can be expressed by the number and effectiveness of pest control species.

The monetary value of selected ES is most often mentioned for pollination because it is easiest one to express on the basis of farmers' sales for the most economically important species, or by the value of produced honey and so on. In case of biodiversity protection, the social value of protected species and habitats of national or European importance, including priority habitats, is established by Nature and Landscape Protection Act (National Council of the Slovak Republic 2002). To express the monetary value of pest and disease control, the following methods were introduced: preventive cost methods (value derived from research of costs to prevent or reduce environmental risk) or a production approach assessment based on indirect loss values which could be caused by pests and diseases on agricultural production.

The flow of supporting ES is also influenced by environmental stress factors, in the form of direct spatial loss of ecosystems, poor/disturbed ecosystem status or the intensity and way of use of surrounding areas. The effectiveness of environmental ES can be enhanced by supporting the proper management of landscape diversity or aiming at increasing the share of ecologically important elements.

The overall capacity of the landscape to provide regulatory and supporting ES is shown in Fig. 6.4. It was expressed as the average value achieved for all ten ES which create this group. The spatial projection expresses the above-mentioned main factors and the context of the provision of these ES. The high value of the landscape capacity is evident in the case of mostly forested mountain and foothill areas, and the low capacity is evident for lowland and basin areas with the predominant arable land. The most important natural regions providing regulatory and supporting ES include the Malé Karpaty, Biele Karpaty, Považský Inovec, Strážovské vrchy, Tríbeč, Vtáčnik, Štiavnické vrchy, Malá Fatra, Veľká Fatra, part of Slovenské Rudohorie, Slovenský kras, Čergov, Slanské vrchy and Východné Karpaty regions.

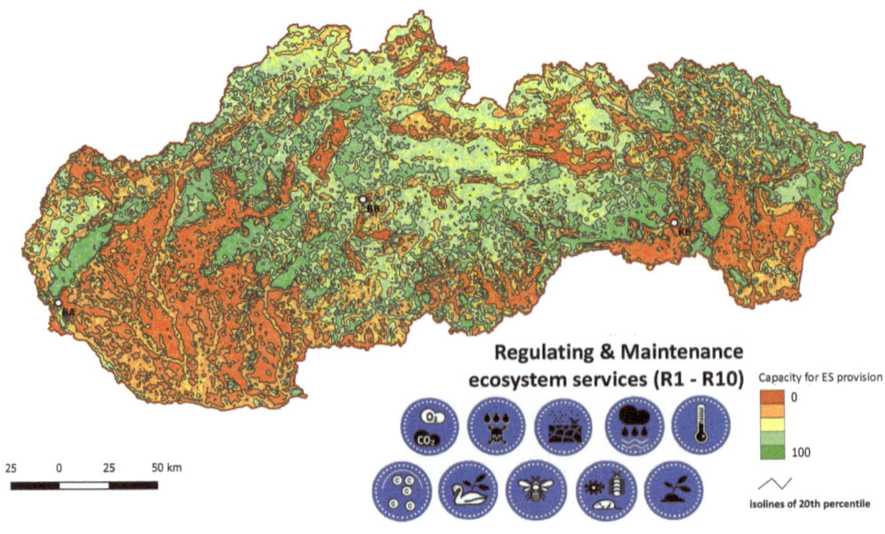

Fig. 6.4 The total capacity of the landscape to provide regulatory and supporting ES

Other mountain ranges and sub-mountain areas are characterized by medium to relatively high landscape capacity. From the lowland and basin areas, the Borská nížina, the peripheral parts of the higher intra-mountain basins, Latorica and Podunajsko areas are the most important in terms of provision of this group of ES.

6.3 Cultural Ecosystems Services

The Millennium Ecosystem Assessment (MEA 2005) defined cultural ES as the intangible benefits which people obtain from ecosystems through spiritual enrichment, cognitive development, reflection, exploration, ability to distinguish values, recreation and aesthetic experiences. These include educational values because the ecosystems, their components and elements, as well as cultural landscapes, form the basis for both formal and informal education, edification and promoting the environmental and cultural-historical awareness as well as shaping the attitudes of the population towards their environment. Inspiration is also beneficial, and not only in art, folklore, but also in the landscape and also the perception of the genius loci. Last but not least, this all presents opportunities for scientific discoveries, research, knowledge and then education and training based on the traditional knowledge of the environment. All the ES and especially the cultural ones often have the character of public goods. This includes conditions of non-exclusion (which means that people cannot be denied ES benefits) and noncompetitive consumption (which means that ES benefits to one person do not reduce their availability to others) (Vačkář et al. 2014).

The cultural ES group includes different ES depending on the classification used (see Table 1.1 in Chap. 1). The most common ones, according to MEA (2005), include the following: recreation and tourism; aesthetic values; cultural diversity; spiritual and religious values; and cognitive and educational values. Three ES were selected for the pilot assessment of cultural ES in Slovakia – recreation and tourism; landscape aesthetics; and natural and cultural heritage.

The general feature of cultural ES is their intangibility and subjectivity. The subjective factors play an important role in the choice of the use of cultural ES. The preference for different services depends largely on the preference of individual residents and visitors, from their values, as well as their social status. The use of intellectual services, especially research and education, is significantly determined by education and employment. These services are preferably used by researchers, educators, nature conservationists, etc., who carry out research activities in individual territories. However, educational services are used by a wider group. In addition to research and teaching staff, they are also used by nature conservationists but also by amateurs interested in knowing the secrets of the landscape, its components and elements. A special group of using cultural ES, especially cultural diversity, includes artists who find inspiration for different types of art in the landscape and its ecosystems. People who find spiritual experiences in the landscape and its ecosystems also form a specific group. Mostly the use of these ES is also conditioned by the occurrence of sacred buildings in the given area (chapels, crosses, churches and other places of worship).

The most commonly used group of the cultural ES includes the services providing physical and experiential interactions, enabling the development of recreation and tourism. These require the presence of special types of ecosystems (especially water and forest ecosystems, agroecosystems as well as urban ecosystems) or special types of landscapes (valuable natural or cultural-historical landscape types). The type of ecosystem often determines the form of recreation and tourism. For example, aquatic ecosystems are most often used for water sports and swimming, with mountain ecosystems being used especially for the development of winter sports – downhill and cross-country skiing, etc., but also for the summer tourism – hiking, forest berries collection and so on. Agro-ecosystems play a primary role in tourism development, and urban ecosystems are dominant in the field of exploration tourism – learning about the cultural and historical values of the landscape. The development of recreation and tourism is also associated with the presence of certain natural resources. The occurrence of mineral and healing waters is tied to the development of the spa industry. A supportive factor for the use of cultural ES is also the socio-economic infrastructure – accommodation, catering facilities, parking lots, educational trails, observation points, cross-country trails, ski resorts, etc. The accessibility of the site and its promotion also play a significant role.

From the point of view of the use of cultural ES, all types of ecosystems and types of the landscape have a certain value, as each of them is specific and requires detailed examination as well as the presentation (research, education-training services, etc.). The most attractive and most desirable include the natural types of ecosystems – sites of protected areas, NATURA 2000 sites, sites with important habitats and others. But they also include cultural and historical landscape features – UNESCO World Natural and Cultural Heritage sites, heritage reserves and zones, traditional landscaping and the like. That is why the positive relationship between the landscape's capacity to provide the ES and significance of the territory in terms of nature and landscape protection is most evident from all ES groups (Fig. 6.5).

However, the real use of many cultural ES is often limited by the need to protect nature, biodiversity and landscape stability as well as the need to protect natural resources (water resources, highest quality soils, forests with special functions and the like). Often this is the source of conflicts between nature protection and various entities benefiting from the use of cultural ES (landowners, operators of accommodation and recreational facilities, etc.). The use of cultural ecosystems is also limited by the effects of stress factors such as environmental contamination (polluted air and water, damaged forest ecosystems, etc.), noise, radiation, localization of inadequate buildings and objects in the landscape and the like.

Due to their intangibility and a high degree of subjectivity, it is relatively difficult to measure, monitor, model and value most of the cultural ES (Schröter et al. 2019). Sociocultural methods are most commonly used to assess them, using participatory methods, such as stakeholder participation workshops, questionnaires, personnel interviews, etc. The main objective of these methods is to identify opinions, demands and attitudes of people in relation to the use of ecosystems (de Groot et al. 2010).

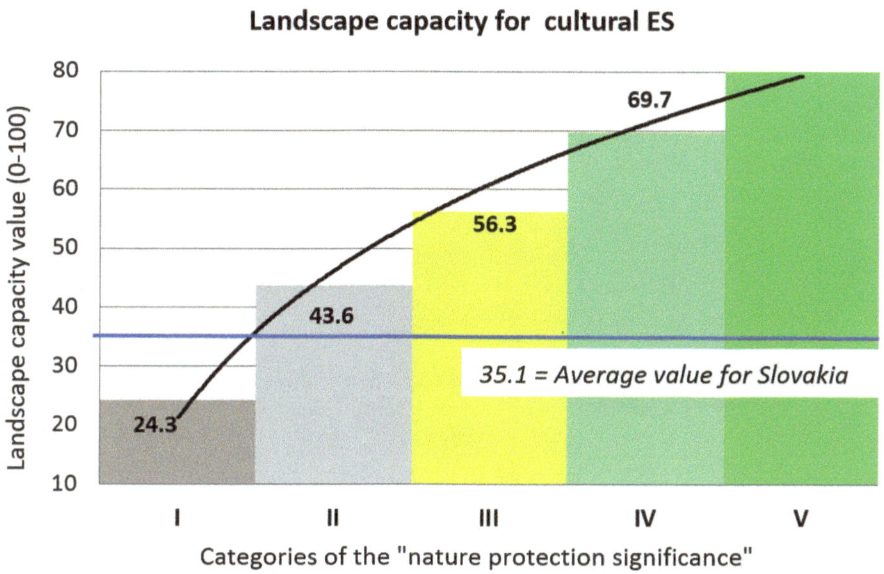

Fig. 6.5 The relationship between cultural ES and the significance of the territory of Slovakia in terms of nature and landscape protection

Photo-series analysis, online map surveys and mobile phone applications are also often used. In the context of the economic assessment of cultural ES, contingent valuation is most often used, which consist in the direct determination of people's willingness to pay or accept compensation for the ES change in a hypothetical market (Farber et al. 2006), e.g. willingness to pay for entrances to protected areas, for revitalization of ecosystems and degraded land, travel costs methods (real consumer costs associated with accommodation, meals, transportation, entrance fees) and so on. The so-called matrix method (Burkhard et al. 2014) and other mapping methods based on the use of GIS and modelling (e.g. ESTIMAP method – Zulian et al. 2013, 2018) are also used.

Various statistical methods are also used for the assessment of the real use of cultural ES (e.g. number of overnight stays, number of tickets sold, number of hunting and fishing permits, number of sports equipment rented, etc.). Groups of cultural ES providing experiential interactions, inspiration for culture and art, spiritual experiences as well as information for exploration are poorly measurable or almost unmeasurable due to the high proportion of subjectivity – or, e.g. the number of created works (literary, art, scientific, etc.) can be used as an indicator.

The landscape's overall capacity to provide cultural ES is shown in Fig. 6.6 and expressed as the average for the three assessed ES, which constitute this group. It should be noted that all three assessed ES are relatively closely related to each other, and their spatial projection is very similar. Therefore, there is no surprise among the territories with the highest capacity for provision of cultural ES – mainly the high mountains of the Carpathians (especially the Vysoké and Západné Tatry) and also

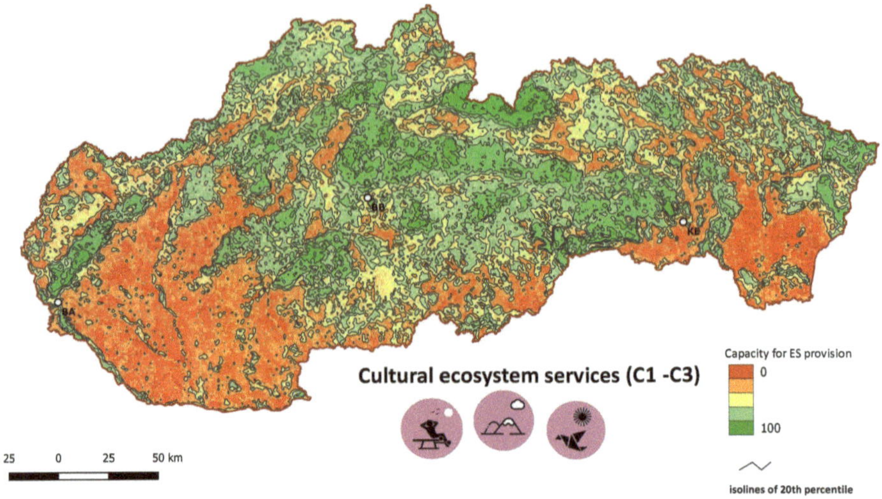

Fig. 6.6 The total capacity of the landscape to provide cultural ES

the areas of Poloniny, the Slovenský raj, Muránska planina, Slovenský kras, Poľana, Štiavnické vrchy and Malé Karpaty. Most of the other mountainous and sub-mountain areas, the Podunajsko and Latorica areas, have medium to high capacity levels. On the contrary, low values are achieved by most of the lowlands and central parts of the intra-mountain basins, with the lowest value being typical for large open lowland and basin highlands.

6.4 Summary Assessment

The main objective of the publication *Catalogue of Ecosystem Services of Slovakia* is to introduce and assess the main ES important for the territory of Slovakia. Of course, staying only at individual ES level without the assessment of groups and the ES as a whole would be insufficient. That is why we decided to prepare a map of the landscape's overall capacity to provide the ES, which represents the synthesis of the first stage of research.

In similar calculations of aggregate indices, the main issue is always to determine the weight (importance) of individual input indicators. In case of the landscape's overall capacity to provide the ES, we have decided to solve this problem relatively simply, but in our opinion in a sufficiently representative and *fair* manner – the weight of provisioning ES as a whole represents 25%, as well as the weight of cultural ES, and finally the weight of regulatory and supporting ES represents 50% of total value. The resulting value was thus calculated as the sum of the capacity values for each ES group multiplied by the given weight. The theoretical value of the capacity ranges from 0 to 100, where 0 means no capacity and 100 the maximum possible landscape capacity for ES provision.

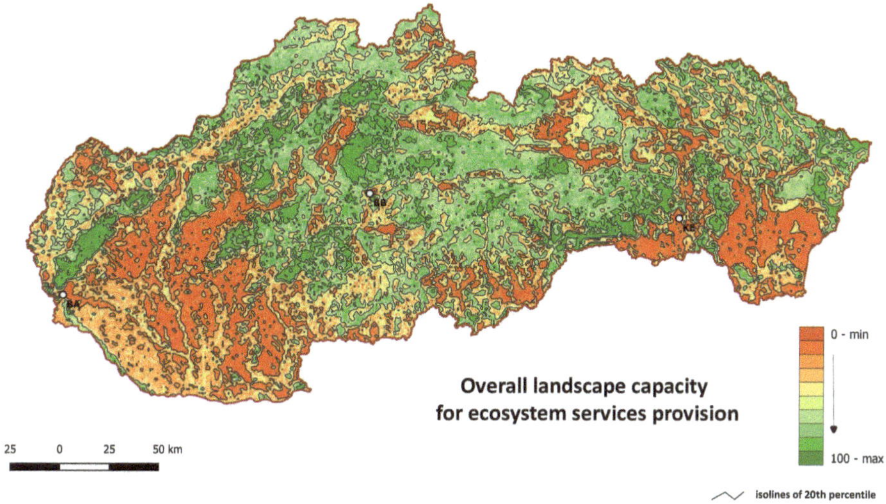

Fig. 6.7 The total capacity of the landscape to provide ecosystem services

The spatial projection of the total landscape capacity is shown in Fig. 6.7. The most important areas of Slovakia from the point of view of ES provision can be quite clearly identified from the map – these include mainly the lower and middle mountain ranges of Slovakia (the largest areas of high landscape capacity are, e.g. in the regions Veľká Fatra, Malá Fatra, Strážovské vrchy, Malé Karpaty, Tríbeč, Vtáčnik, Štiavnické vrchy, Poľana, Starohorské vrchy, Slovenský raj, Muránska planina, Slovenský kras, Slanské vrchy and Bukovské vrchy). On the contrary, the lowest capacity of the landscape is typical for larger areas of lowlands and basins of Slovakia (Chvojnická pahorkatina, Podunajská nížina, Juhoslovenská kotlina, Košická kotlina, Východoslovenská nížina), of the smaller intra-mountain basins, these include Turčianska kotlina, Oravská kotlina, Hornádska kotlina and Žiarska kotlina.

As ecosystem functions and related services are substantially based on the natural structure of the territory, it also seems useful to assess the main natural units in terms of their ES capacity. The main natural regions of Slovakia are well represented by geomorphological units (GM). Table 6.1 shows the average capacity values of these units for the provision of ES – both the total values and the values for the ES main groups. Colour highlighting of values in table cells expresses the division of units based on so-called 20 percentile (every 20% of the total number of GM units is represented by a different colour – the best 20% is highlighted in dark green and the worst 20% in orange). The total values are shown in Fig. 6.8.

Based on these values, it is possible to consider the Spišsko-gemerský kras (Slovenský raj a Muránska planina), Veľká Fatra, Malá Fatra, Slovenský kras, Bukovské vrchy, Starohorské vrchy, Strážovské vrchy, Čergov, Malé Karpaty and Slanské vrchy as the *ten most important GM in Slovakia*. In general, almost all GM from the top 20 of the most important show *high and very high capacity values for*

Table 6.1 Average values of the landscape's capacity to provide ES for geomorphological units in Slovakia

Number	Geomorphological unit	Geomorphological region	km^2	TOT_ ES	PROV_ ES	REG_ ES	CULT_ ES
1	Spišsko-gemerský kras	Slovenské Rudohorie	364	58.56	46.87	58.73	69.91
2	Veľká Fatra	Fatransko-tatranská oblasť	786	57.59	46.87	57.56	68.36
3	Malá Fatra	Fatransko-tatranská oblasť	550	56.23	42.41	58.72	65.09
4	Slovenský kras	Slovenské Rudohorie	496	54.81	43.52	55.52	64.70
5	Bukovské vrchy	Poloniny	378	54.80	29.57	61.55	66.58
6	Starohorské vrchy	Fatransko-tatranská oblasť	177	54.54	45.12	55.63	61.77
7	Strážovské vrchy	Fatransko-tatranská oblasť	960	54.44	48.91	57.95	52.96
8	Čergov	Východné Beskydy	310	54.32	38.85	60.41	57.61
9	Malé Karpaty	Fatransko-tatranská oblasť	848	54.03	43.24	56.83	59.24
10	Slanské vrchy	Matransko-slanská oblasť	525	53.01	39.06	60.75	51.44
11	Tribeč	Fatransko-tatranská oblasť	506	52.97	41.97	59.43	51.05
12	Burda	Matransko-slanská oblasť	28	52.96	39.53	57.74	56.69
13	Moravsko-sliezske Beskydy	Západné Beskydy	24	52.90	56.35	48.79	57.81
14	Poľana	Slovenské stredohorie	181	52.83	35.39	53.60	68.73
15	Vtáčnik	Slovenské stredohorie	365	52.62	37.79	59.69	53.32
16	Nízke Tatry	Fatransko-tatranská oblasť	1268	52.55	40.68	51.12	67.29
17	Vihorlatské vrchy	Vihorlatsko-gutinská oblasť	382	52.35	36.53	59.45	54.07
18	Štiavnické vrchy	Slovenské stredohorie	871	52.29	37.45	56.01	59.69
19	Čierna hora	Slovenské Rudohorie	264	52.11	38.34	57.48	55.12
20	Volovské vrchy	Slovenské Rudohorie	1352	51.94	33.86	56.37	61.17
21	Tatry	Fatransko-tatranská oblasť	543	51.87	28.72	44.97	88.97
22	Pohronský Inovec	Slovenské stredohorie	153	51.59	39.92	59.24	47.98
23	Považský Inovec	Fatransko-tatranská oblasť	465	50.25	41.54	57.72	44.04

(continued)

Table 6.1 (continued)

Number	Geomorphological unit	Geomorphological region	km²	TOT_ ES	PROV_ ES	REG_ ES	CULT_ ES
24	Súľovské vrchy	Fatransko-tatranská oblasť	194	50.15	48.86	51.95	47.84
25	Busov	Nízke Beskydy	99	49.51	45.51	55.06	42.37
26	Chočské vrchy	Fatransko-tatranská oblasť	117	49.49	42.41	49.70	56.15
27	Stolické vrchy	Slovenské Rudohorie	603	47.60	31.61	54.67	49.46
28	Kremnické vrchy	Slovenské stredohorie	485	47.21	42.53	51.09	44.14
29	Oravské Beskydy	Stredné Beskydy	139	46.88	39.48	46.32	55.40
30	Laborecká vrchovina	Nízke Beskydy	1158	46.83	34.86	54.52	43.44
31	Javorníky	Slovensko-morav. Karpaty	867	46.60	41.70	48.35	47.99
32	Branisko	Fatransko-tatranská oblasť	84	46.43	33.84	52.46	46.95
33	Žiar	Fatransko-tatranská oblasť	146	46.40	42.55	51.33	40.39
34	Veporské vrchy	Slovenské Rudohorie	898	46.05	34.53	49.45	50.75
35	Javorie	Slovenské stredohorie	229	45.90	37.84	50.50	44.74
36	Kysucké Beskydy	Stredné Beskydy	168	45.20	37.01	46.10	51.62
37	Oravská Magura	Stredné Beskydy	173	45.04	35.50	48.00	48.67
38	Biele Karpaty	Slovensko-morav. Karpaty	681	43.94	33.57	48.49	45.20
39	Kysucká vrchovina	Stredné Beskydy	418	43.78	36.24	45.30	48.28
40	Kozie chrbty	Fatransko-tatranská oblasť	170	43.63	44.15	43.72	42.92
41	Ostrôžky	Slovenské stredohorie	259	42.84	32.53	50.59	37.64
42	Turzovská vrchovina	Západné Beskydy	223	42.79	48.05	39.72	43.66
43	Podtatranská brázda	Podhôľno-magurská oblasť	89	41.86	32.16	40.41	54.45
44	Revúcka vrchovina	Slovenské Rudohorie	949	41.80	30.81	48.70	38.98
45	Ľubovnianska vrchovina	Východné Beskydy	189	40.72	30.70	47.38	37.31
46	Cerová vrchovina	Matransko-slanská oblasť	500	39.78	30.93	47.02	34.26
47	Spišská Magura	Podhôľno-magurská oblasť	344	38.53	27.61	41.48	43.52
48	Levočské vrchy	Podhôľno-magurská oblasť	644	38.52	29.13	42.81	39.32

(continued)

Table 6.1 (continued)

Number	Geomorphological unit	Geomorphological region	km²	TOT_ES	PROV_ES	REG_ES	CULT_ES
49	Skorušinské vrchy	Podhôľno-magurská oblasť	192	38.19	34.83	40.63	36.68
50	Bachureň	Podhôľno-magurská oblasť	130	37.93	32.81	43.06	32.79
51	Krupinská planina	Slovenské stredohorie	856	37.84	33.66	42.79	32.12
52	Zemplínske vrchy	Matransko-slanská oblasť	109	37.11	33.93	41.52	31.46
53	Ondavská vrchovina	Nízke Beskydy	1807	37.00	38.23	41.16	27.45
54	Bodvianska pahorkatina	Lučensko-košická zníženina	155	36.65	36.18	40.59	29.57
55	Pieniny	Východné Beskydy	53	36.38	22.11	41.14	41.78
56	Horehronské podolie	Fatransko-tatranská oblasť	316	35.22	30.73	35.79	38.58
57	Podbeskydská vrchovina	Stredné Beskydy	235	34.63	28.79	35.30	39.13
58	Borská nížina	Záhorská nížina	1162	34.52	36.18	38.26	25.40
59	Šarišská vrchovina	Podhôľno-magurská oblasť	274	33.19	28.38	37.23	29.93
60	Beskydské predhorie	Nízke Beskydy	671	32.30	27.71	37.50	26.55
61	Jablunovské medzihorie	Západné Beskydy	53	32.03	29.75	34.77	28.85
62	Dolnomoravský úval	Juhomoravská panva	97	31.32	39.87	31.49	22.44
63	Zvolenská kotlina	Slovenské stredohorie	625	31.23	31.64	31.58	30.12
64	Oravská vrchovina	Stredné Beskydy	284	31.06	19.25	36.16	32.68
65	Rožňavská kotlina	Slovenské Rudohorie	67	29.51	30.92	28.44	30.25
66	Hornonitrianska kotlina	Fatransko-tatranská oblasť	400	29.45	33.86	30.25	23.45
67	Myjavská pahorkatina	Slovensko-morav. Karpaty	365	29.14	29.55	30.25	26.50
68	Považské podolie	Slovensko-morav. Karpaty	561	28.96	37.67	28.43	21.31
69	Žilinská kotlina	Fatransko-tatranská oblasť	271	28.63	31.85	27.65	27.37
70	Podtatranská kotlina	Fatransko-tatranská oblasť	1197	28.01	25.29	26.57	33.61
71	Podbeskydská brázda	Stredné Beskydy	132	26.13	25.01	24.87	29.78
72	Spišsko-šarišské medzihorie	Podhôľno-magurská oblasť	513	25.99	24.03	28.25	23.41

(continued)

Table 6.1 (continued)

Number	Geomorphological unit	Geomorphological region	km²	TOT_ ES	PROV_ ES	REG_ ES	CULT_ ES
73	Pliešovská kotlina	Slovenské stredohorie	100	25.88	34.11	24.45	20.53
74	Juhoslovenská kotlina	Lučensko-košická zníženina	1805	25.88	29.91	27.81	18.07
75	Žiarska kotlina	Slovenské stredohorie	128	25.09	28.11	26.43	19.38
76	Oravská kotlina	Podhôľno-magurská oblasť	216	24.71	21.35	24.70	28.14
77	Východoslovenská pahorkatina	Východoslovenská nížina	718	24.23	28.40	25.31	17.94
78	Podunajská rovina	Podunajská nížina	3458	23.92	36.68	22.57	13.88
79	Košická kotlina	Lučensko-košická zníženina	1141	23.84	28.07	24.45	18.43
80	Turčianska kotlina	Fatransko-tatranská oblasť	436	23.84	30.17	22.36	20.44
81	Hornádska kotlina	Fatransko-tatranská oblasť	462	23.00	26.43	21.82	21.93
82	Chvojnícka pahorkatina	Záhorská nížina	353	22.94	30.58	22.85	15.49
83	Podunajská pahorkatina	Podunajská nížina	6355	21.83	29.37	21.92	14.13
84	Východoslovenská rovina	Východoslovenská nížina	1716	19.65	27.49	20.52	10.07
Slovak Republic			49.035	36.80	34.20	38.90	35.10

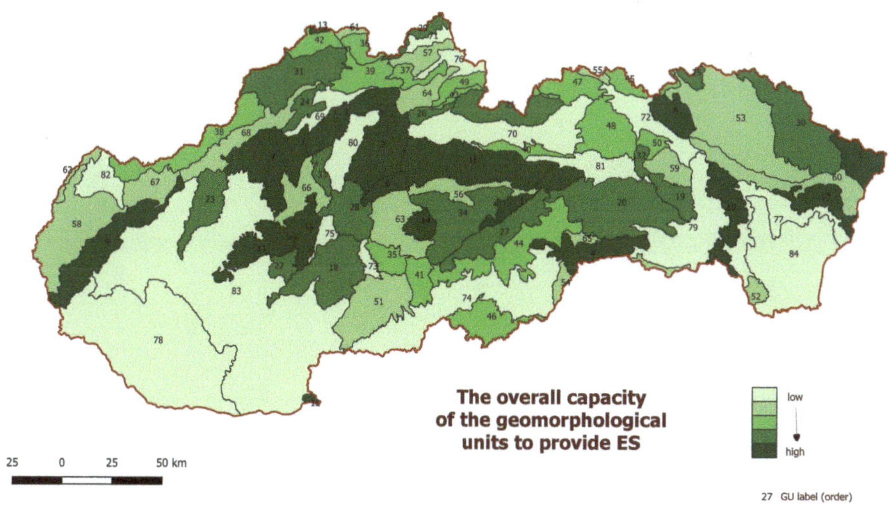

Fig. 6.8 The total capacity of the landscape to provide ES for geomorphological units of Slovakia

all ES groups (with the exception of Bukovské vrchy, Poľana and Volovské vrchy, which have only average provisioning ES capacity).

On the contrary, the least significant GM in Slovakia in terms of capacity to provide the ES can be considered the areas of Východoslovenská rovina, Podunajská pahorkatina, Chvojnícka pahorkatina, Hornádska kotlina, Turčianska kotlina, Košická kotlina, Podunajská rovina, Východoslovenská pahorkatina and Oravská kotlina a Žiarska kotlina. These units, together with other predominantly intra-mountain basin areas, have very low capacity for regulatory and cultural ES and predominantly low capacity for provisioning ES. The exception is the GM of Podunajská rovina with a high value for provisioning ES (mainly due to high capacity for ES P1, P3 and P4).

An interesting indicator of the balance of GM in terms of ES provisioning capacity is also the difference between the most favourable and least favourable value, which is the lowest in the case of Súľovské vrchy, Kozie chrbty (with mostly high value of the landscape's capacity), Zvolenská kotlina and Rožňavská kotlina (with low landscape capacity). On the contrary, the largest difference is present in case of GM units of Bukovské vrchy, Poľana, Nízke Tatry, Volovské vrchy and especially the Tatra Mountains – in all cases the most favourable values are achieved for the cultural ES and least favourable for the provisioning ES.

Figure 6.9 shows the relationship between landscape structure and its capacity to provide ES. As the land use has directly entered the computational algorithms for most ES, such an assessment is merely a summary of how individual categories of

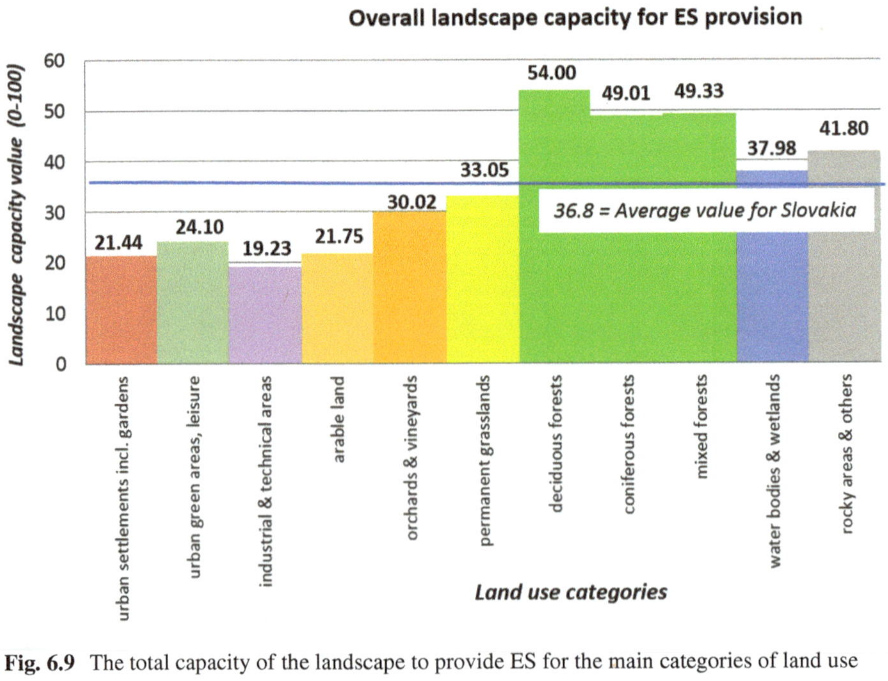

Fig. 6.9 The total capacity of the landscape to provide ES for the main categories of land use

the landscape structure contribute to ES provision. Statistical results confirm the generally accepted fact that forest ecosystems are the most important type of ecosystems in terms of ES provision and urbanized areas are the least important.

The most important category of landscape structure in terms of ES provision in Slovakia is the *deciduous forests* with the highest value of the landscape's capacity (it has 54 points, which means 1.5 times the average value of the whole territory of Slovakia). *Mixed and coniferous forests* are also among the most important categories. The second group of significance with values above the national average includes two types of landscape structure: *rocks and screes* and *water areas and wetlands*. While the first type is particularly significant due to the very high value for cultural ES and achieves a very small spatial extent, hydric ecosystems are quite significant from the point of view of all three ES groups, but most for the regulatory ES.

Grassland (meadows and pastures) and *permanent agricultural crops* (orchards and vineyards) have an average significance in terms of ES provision in Slovakia. Their capacity to provide individual ES groups is relatively balanced.

Other major categories of the landscape structure – arable land and urbanized areas – have low to very low significance in terms of ES provision in Slovakia. *Arable land* has a higher capacity for provisioning services; *residential vegetation, sports and recreational areas* provide relatively balanced, albeit lower potential for all major ES groups. The lowest overall significance in terms of capacity for the provision of ES comes from *residential buildings* and, in particular, *industrial and technical areas*.

The assessment of the landscape's capacity to provide ES is only the first step of a comprehensive ES assessment. As reported by Burkhard et al. (2014), in the ES assessment, it is necessary to distinguish three basic aspects – from the landscape's potential to provide the ES (supply, capacity) through the requirements for their provision in a particular territory (demand) to their real use and balance (ES flow).

Landscape capacity refers to the usable potential of natural resources and ecosystem functions. It creates the so-called ES supply, which according to Burkhard et al. (2014) is based on potentials and additional inputs. These inputs are related to the economy and represent social, human, financial and production investment assets (Costanza and Daly 1992). The ES flow is realized between ES supply and consumption, reflecting the real amount of man-made goods and ecosystem services in a particular territory (in the form of a vector from production sites to consumption points), thus generating the final benefit from ES to humans. This flow is directed from the natural environment to human society and determined by the so-called *ES demand* in a particular territory and over a period of time (Burkhard et al. 2014).

ES supply, demand and flow together create a dynamic process of creating and using the ES which moves from natural ecosystems to human society – a simplified scheme of this process is shown in Fig. 6.10.

Only when all the basic aspects of ES provision and use in Slovakia are known and assessed, we can state that there is a *comprehensive ES assessment in Slovakia*. The present publication is therefore only one of the necessary parts of such an

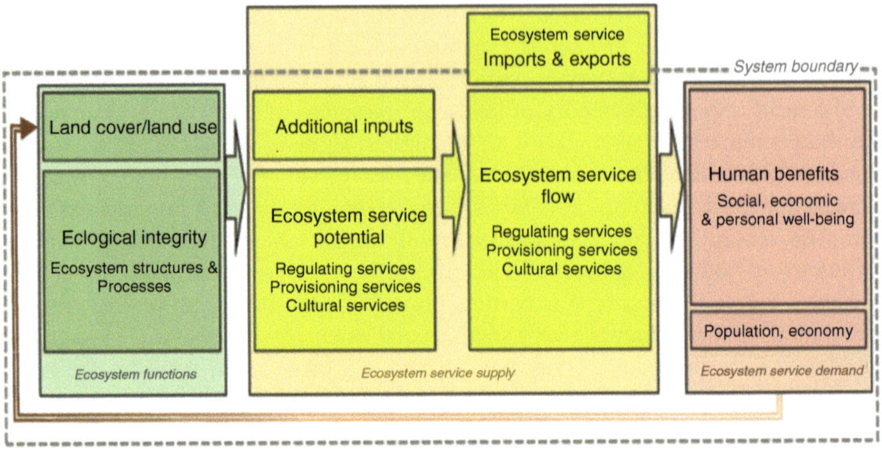

Fig. 6.10 The ecosystem services cycle in the landscape and society – conceptual model. (Source: Burkhard et al. 2014)

assessment. The individual ES chapters also provide appropriate data sources for assessing the demand and use of these ES, but only theoretically – the specific methodologies will depend on the data actually available. However, this process goes well beyond the environmental sector and will require much better synergies from other sectors – in particular, the availability of selected data sources. It, therefore, remains a challenge for the coming future.

6.5 The Importance of Ecosystem Services for Nature and Landscape Protection in Slovakia

The welfare in EU countries is supported by its natural capital, which includes ecosystems providing basic goods and services to people. Publications on the economic benefits of the NATURA 2000 network (2013) show that NATURA 2000 plays a key role in protecting and strengthening the EU's natural capital. NATURA 2000 network has an important role to play in addressing the challenges in relation to climate change by mitigating the changes and impacts. It also includes carbon-rich habitats, brings socioeconomic benefits such as maintaining the water cycle and its quality, preserves natural pollinators, preserves landscape values and promotes tourism and recreation and the like. According to this study, the benefits flowing from NATURA 2000 are on the order of 200–300 billion EUR per year. NATURA 2000 network can be seen as a key element of green infrastructure in an open landscape, involved or directly providing a number of ecosystem services which are threatened by the degradation of natural habitats. Investing in NATURA 2000 management and recovery measures and strong legal protection can increase the provision of these services.

We have tried to document the significance of the ES in terms of nature and landscape protection (or the relationship between these two aspects of land use and management) by comparing the landscape's capacity to provide the ES and the so-called significance of the territory in terms of nature and landscape protection (for more details see Chap. 2). The achieved results (presented and commented in the subchapters of this publication) show a clear correlation between these two indices in the case of the majority of the ES, which is particularly evident in the case of cultural ES and most regulatory ES (except for R4 and R10).

Nature protection plays an indispensable role, particularly in the provision of regulatory/supporting and cultural ES. The greatest capacity to provide regulatory and supporting ES comes from natural and seminatural ecosystems, which are also most significant in terms of NaLP – the most obvious positive correlation has been documented for ES R1, R3, R7 and R9. Also in the case of cultural ES, the natural ecosystems and significant cultural and historical landscape features are the most attractive – a very clear positive correlation with the significance of the territory in terms of nature and landscape protection was recorded for all three ES (C1–C3).

A slightly different picture applies to the provisioning ES, for which the landscape's capacity to provide ES as a whole is least related to nature protection. Actually, in the case of ES P1, we recorded a negative correlation and in the case of ES P2 a neutral relationship. However, as a whole, also here is a slightly positive relationship between these two indices.

The relationship between the overall capacities of the landscape to provide the ES (see Fig. 6.7) and the significance of the territory in terms of nature and landscape protection (degrees 1–5) is shown in Fig. 6.11. A positive correlation is more

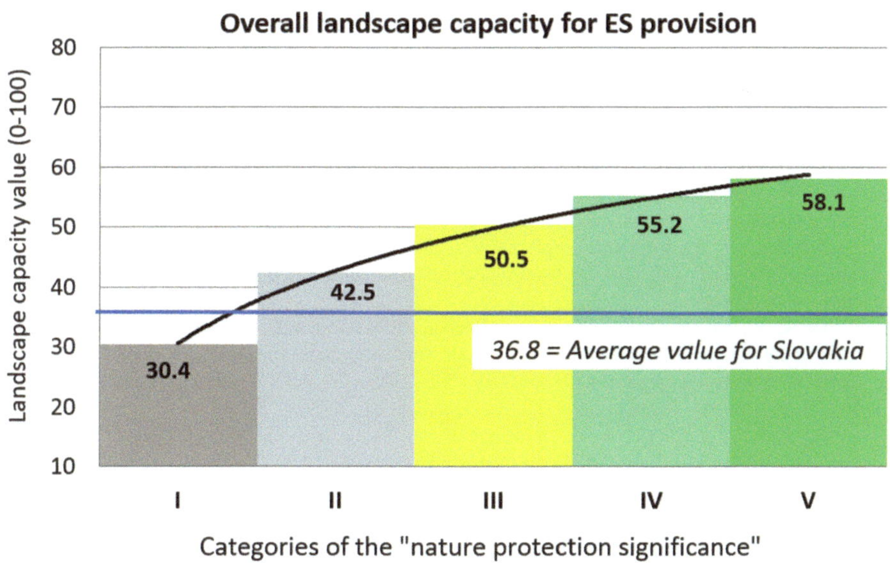

Fig. 6.11 The overall relationship between the landscape capacity to provide ES and the significance of the territory of Slovakia in terms of nature and landscape protection

than evident – the value of the landscape's capacity increases with the significance of the territory in terms of NaLP, which applies to all categories of significance. This fact emphasizes the key importance of nature and landscape protection not only for the healthy state of ecosystems but also for the fulfilment of their functions and services that are directly or indirectly used by humans.

The ES concept *fundamentally changes the view of nature protection functions and tasks*. While the approaches to date have focused mainly on the subjects of protection (and thus habitats, species and biodiversity), the ES concept brings a different view of the mission and role of protected areas, especially through the protection of processes and related ecosystem functions. Thus, the ES concept requires a change in the traditional *protection paradigm* – the most important ES producers are natural ecosystems and habitats – even those which are relatively widespread. The rarity and the level of endangering of the habitats in this concept are diminishing, and the existence and presence of habitats in places where there is a demand for the relevant ES play the most important role. The accessibility of the ES then also plays an important role, with the ES best being accessible in the largest possible area with benefits of ES provided to as many people as possible – ideally as close as possible to the demand sites (i.e. the occurrence of the largest number of residents or visitors in a particular territory).

Only time will tell, whether this approach is correct and whether it is realistic and useful to change the long-term nature and landscape protection *strategy*. However, it is clear that it is more than necessary to invest resources in the preservation and improvement of protected areas in Slovakia. The assessment clearly showed that the protected nature areas provide most of the natural services and benefits which a man uses directly or indirectly – most of the ES are associated with those parts of Slovakia where protected areas are most represented.

References

Burkhard, B., Kandziora, M., Hou, Y., & Müller, F. (2014). Ecosystem service potentials, flows and demands – Concepts for spatial localisation, indication and quantification. *Landscape Online, 34*(1), 1–32. https://doi.org/10.3097/LO.201434.

Costanza, R., & Daly, H. (1992). Natural capital and sustainable development. *Conservation Biology, 6*, 37–46.

De Groot, R. S., Alkemade, R., Braat, L., Hein, L., & Willemen, L. (2010). Challenges in integrating the concept of ecosystem services and values in landscape planning, management and decision making. *Ecological Complexity, 7*, 260–272.

Farber, S., Costanza, R., Childers, D. L., Erickson, J., Gross, K., Grove, M., Hopkinson, C. S., Kahn, J., Pincetl, S., Troy, A., Warren, P., & Wilson, M. (2006). Linking ecology and economics for ecosystem management. *BioScience, 56*(2), 121–133. https://doi.org/10.1641/0006-3568(2006)056[0121:LEAEFE]2.0.CO;2.

Millennium Ecosystem Assessment. (2005). *Ecosystems and human well-being: Wetlands and water synthesis*. Washington, DC: World Resources Institute.

National Council of the Slovak Republic – NR SR. (2002). Zákon NR SR 543/2002 Z. z. o ochrane prírody a krajiny v znení neskorších predpisov.

Schröter, M., Bonn, A., Klotz, S., Seppelt, R., & Baessler, C. (2019). *Atlas of ecosystem services. Drivers, risks, and societal responses*. Cham: Springer.

The Economic Benefits of the Natura 2000 Network. (2013). Luxembourg: Publications Office of the European Union.

Vačkář, D., Frélichová, J., Lorencová, E., Pártl, A., Harmáčková, Z., & Loučková, B. (2014). *Metodologický rámec integrovaného hodnocení ekosystémových služeb v České republice*. http://www.ecosystemservices.cz/cs/metodologicky-ramec-integrovaneho-hodnoceni-ekosystemovych-sluzeb-v-ceske-republice. Accessed 13 Apr 2019.

Zulian, G., et al. (2013). *ESTIMAP: Ecosystem services mapping at European scale*. Brussels: Institute for Environment and Sustainability, Joint Research Centre, European Commission.

Zulian, G., Stange, E., Woods, H., Carvalho, L., Dick, J., Andrews, C. H., Baró, F., Vizcaino, P., Barton, D. N., Nowel, M., Rush, G. M., Autunes, P., Fernandes, J., Ferraz, D., Ferreira Dos Santos, R., Aszalós, R., Aranay, I., Crúcz, B., Priess, J. A., Hoyer, C. H., Bürger-Patricio, G., Lapola, D., Mederly, P., Halabuk, A., Bezák, P., Kopperoinen, L., & Viinikka, A. (2018). Practical application of spatial ecosystem service models to aid decision support. *Ecosystem Services, 29*(C), 465–480. https://doi.org/10.1016/j.ecoser.2017.11.005.

Conclusion

Peter Mederly and Ján Černecký

The results presented in this publication can be considered as the completion of the first stage of ES assessment in Slovakia. Its objective was to provide a transparent assessment of the ES important for the territory of Slovakia – especially the basic characteristics of the ES and the assessment of the landscape's capacity to provide them. At the beginning of the publication, the basics of ES concept theory (basic terminology, ES classification, and basic assessment methods) and the progress of the application of this concept in Europe and Slovakia are presented in a clear form. The main part of the publication is focused on the ES assessment for the territory of Slovakia – a total of 18 ES were selected, which are divided into three main groups (provisioning, regulatory/supporting, and cultural ES). For each ES, its brief characteristics, used methods of assessment, the main types of ecosystems important for its provisioning, significance from the point of view of nature protection, and the specific assessment of the given ES for the territory of Slovakia are presented. The final part of the publication includes the summarized results – first for the three main groups of the ES and finally the overall ES assessment for the territory of Slovakia.

Of course, we do not consider the ES assessment process in Slovakia to be completed. As stated in the final assessment, the next stage should include the assessment of other aspects of ES provision – their real flow in the landscape, the demand for individual ES, as well as the economic (monetary) assessment of the benefits

P. Mederly (✉)
Faculty of Natural Sciences, Constantine the Philosopher University in Nitra, Nitra, Slovakia
e-mail: pmederly@ukf.sk

J. Černecký
Faculty of Natural Sciences, Constantine the Philosopher University in Nitra, Nitra, Slovakia

State Nature Conservancy of the Slovak Republic, Banská Bystrica, Slovakia

Institute of Landscape Ecology of the Slovak Academy of Sciences, Bratislava, Slovakia

P. Mederly, J. Černecký (eds.), *A Catalogue of Ecosystem Services in Slovakia*,
https://doi.org/10.1007/978-3-030-46508-7

which ecosystems provide. However, not all agree with the need for an economic assessment of ecosystems and their functions – some scientists do not agree with it for various reasons, most often to reduce the wide range of values to solely monetary units. Economic ES assessment is relatively poorly developed in Slovakia. Although some examples of such studies for the protected areas, forest ecosystems, soils, or water exist, a comprehensive and aggregate monetary ES assessment is still absent. Therefore, Slovakia is in this respect far behind other European countries (Finland, Flanders, the Netherlands, Germany, Romania, Spain, the United Kingdom, Czech Republic).

The results of monetary valuation show that nature protection not only fulfils the function of preserving natural values for future generations but obviously also saves costs in the future and thus contributes significantly to the economic domain. Therefore, it can be unequivocally stated that the system of protected areas in Slovakia is a good economic investment for the future – it is reasonable to assume that the value of ecosystems in protected areas will continue to grow with effective and proper nature protection. In terms of nature protection in Slovakia, the ES concept is relatively new and is still not sufficiently implemented in this area. The basic framework is provided by the Nature Conservancy Act which was the first to define the ES at the national level and thus provide an initial legislative anchor. The ES concept is also incorporated in the Updated National Strategy for Biodiversity Protection 2012–2020 and in the Environmental Policy Strategy of the Slovak Republic 2030. It is therefore evident that the ES concept is gradually being introduced in Slovakia but has so far been underpinned mainly by the international commitments. That is why it is necessary to continue to develop it – not only within the framework of nature and landscape protection but also in decision-making on landscape management, spatial and territorial planning, and environmental impact assessment. Financial and moral investment in this area is certainly a wise choice but also very economically beneficial. It is up to the current generation whether and to what extent wise decisions will contribute to improving the quality of life in Slovakia.

FSC
www.fsc.org
MIX
Papier | Fördert gute Waldnutzung
FSC® C083411

Kolibri 360 GmbH
Ferdinand-Jühlke-Straße 7
99095 Erfurt, Deutschland
produktsicherheit@kolibri360.de